HEARTBREAK

Also By Florence Williams

The Nature Fix:
Why Nature Makes Us Happier, Healthier,
and More Creative

Breasts:
A Natural and Unnatural History

HEARTBREAK

A Personal and
Scientific Journey

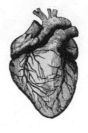

Florence Williams

W. W. NORTON & COMPANY
Independent Publishers Since 1923

Heartbreak is a work of nonfiction. Most of the dialogue is based on taped conversations; as for the rest, the dialogue has been reconstructed based on the author's notes. Some names of individuals have been changed, along with potentially identifying characteristics, and some timelines have been changed. The approaches, techniques, and remedies referred to or discussed in this book reflect the author's experiences and/or opinions only, are not intended as recommendations, and should not be construed as a substitute for medical, therapeutic, or other professional advice.

For information about permission to reproduce selections from this book, write to
Permissions, W. W. Norton & Company, Inc., 500 Fifth Avenue, New York, NY 10110

For information about special discounts for bulk purchases, please contact
W. W. Norton Special Sales at specialsales@wwnorton.com or 800-233-4830

Manufacturing by Lake Book Manufacturing
Book design by Beth Steidle
Production manager: Beth Steidle

ISBN 978-1-324-00348-9

W. W. Norton & Company, Inc., 500 Fifth Avenue, New York, N.Y. 10110
www.wwnorton.com

W. W. Norton & Company Ltd., 15 Carlisle Street, London W1D 3BS

1 2 3 4 5 6 7 8 9 0

In memory of John Cacioppo and Jaak Panksepp,
who taught us that emotions matter, especially
the tricky ones, and in loving memory of Penny
Williams, who told me I was going to be okay.

CONTENTS

PART THREE

AWE

INTRODUCTION

The invention of the ship is also the
invention of the shipwreck.
—JENNY OFFILL, *WEATHER*

My biggest problem at the moment was the portable toilet. It was just too heavy. It was weighing down the bow of my canoe, which was already loaded with 80 pounds of water and a double-walled cooler filled with fairly ridiculous items like coconut milk, rib-eye steaks, and cage-free liquid whole eggs. Also, I'd brought a fetching beach parasol. But why does something you shit in in the desert have to be made of ammunition-grade 20-millimeter steel? It doesn't! I just needed some plastic bags. The ill-conceived toilet was just one of many small and giant mistakes that had led me to this moment, cursing alone in the wilderness. There were the mistakes in my marriage, the cosmic mistake (to my mind) of the divorce, the wrong men I'd fallen for in the year since my separation, the friendships I'd overburdened. All of these were, yes, weighing me down. If I thought about the heavy-shit metaphors too long, my head hurt.

Most recently, there was the poor decision, made because I was possibly having a hot flash, to launch this leg of my journey a day early, at 7 p.m., in fading light, just above a small rapid, in a canoe that felt like it weighed a thousand pounds. Then again, it was August and it was 97 degrees in Green River, Utah. Even a teenage boy would be having a hot flash. Camping at the shadeless town park was an unbearable option. Running a desert river for a month in the height of summer was probably another bad decision. But here I was. An outfitter named Craig had rented me the 15-foot canoe with a broken thwart, splintering gunwales, and the tanker toilet. The boat was the color of lipstick you wear when you're trying too hard. It did, however, match the parasol.

"Just remember," he'd said, "if you don't know knots, make lots!" He laughed, snapped a picture of me surrounded by my gear, and drove off in his air-conditioned pickup.

To be clear, I do know how to tie knots and I generally know what I'm doing in the wilderness. But my own canoe lay upside down in Washington, DC, where it petulantly awaited better days and where, until recently, I also petulantly lay, often right side down, after my husband decided to leave our 25-year marriage because, among other things, he said he needed to go find his soul mate. Still, nothing in my prior canoeing experience had fully prepared me for the reality that I could barely alter the trajectory of this boat once I got it into the river. Only a few small inches of freeboard lay between the water and the top of my gunwales. I stared at the approaching shoals. I glared at the toilet, glinting like a smug brigadier in the twilight.

The river split into two channels. I chose the one on the right, but the current grew fast, shallow, and bumpy. The canoe scraped over some rocks, then some more, and started to list sideways. I pushed my feet back into my river shoes and hopped out into the shin-deep water, figuring I'd have an easier time keeping the boat upright and off the rocks if I were outside it. My heart was beating fast, and I chastised myself for not tying down my gear better. The boat bumped along, upright, and I jumped back in. I knew I needed to pull over and camp, soon, before it got any darker. I grounded the boat onto the first available scruffy gravel bar. For my first night ever spent alone in the wilderness, I'd be camping within sight and earshot of the interstate.

By this point, I'd already been paddling the Green River for two weeks. With friends and family, I'd run Split Mountain and Desolation Canyon, among other stretches. Everything was so aptly named. The first white men to run these canyons 150 years ago hated most of it. "Country worthless," scrawled one in his journal. Now it was just me and the sound of air brakes. I spent the night awake, berating myself for existing in the first place, then berating my husband, and then scheming about how to jettison the toilet, because there was no way I was hauling that thing for the next two weeks.

I was here because I needed to jettison so many things. A year's worth of fear, dismay, and loneliness, some bad habits acquired to stave off those feelings, a peevish and lingering sense of abandonment, my stubborn attachment to a man who was clearly no longer in my boat.

A year and a half earlier, I'd had a life that seemed worth keeping afloat. I was a science journalist with two trusting and kind-hearted teenagers and a husband who ran a venerable and useful nonprofit. We had moved a few years before from Colorado, and I missed our former life out west, but we had a comfortable house near the Potomac River and a goofy mutt to walk the towpath with. I thought our long marriage was fundamentally sound and totally salvageable. We'd seen each other through academic degrees and professional milestones. We'd had decades of fun living in beautiful places, and we'd produced these amazing little people who depended on us for love and hope and stability in a world that seemed to be growing only more confusing and unjust. Our friends' marriages weren't perfect either, but they worked things out. As far as I could tell, nobody here was miserable or violent or crazy or impossibly annoying.

———

HAVING NEVER BEEN heartbroken before, I tended to dismiss portrayals of it in popular culture or literature or even by my friends, I'm sorry to say, as overwrought. But one of the first stages of heartbreak, I soon learned, is feeling stunned, even if you shouldn't have been. I'd been used to feeling in control. But you can't game heartbreak. It overtakes you. When my husband decided to live on his own after three decades of togetherness, the clichés of heartbreak felt not like melodrama at all. I felt like I'd been axed in the heart, like I was missing a limb, set adrift in an ocean, loosed in a terrifying wood. I felt imperiled. Our dyad had dissolved into vapor, and I couldn't grasp what remained. I still plodded through my days, cooking for our son and daughter and walking the dog and making most but not all of my deadlines. I would have moments of collapse, and then I'd get back up feeling vacant and dense at the same time.

We spent another six months still living together and drafting—painfully drafting—a detailed parenting plan. By the time he took his small blue suitcase and rolled it out the kitchen door forever, I'd already lost 20 pounds I didn't want to lose. I couldn't imagine a life without him. Or ever trusting men again, or being able to love or be loved. Having just turned 50, it all seemed even more impossible. I was completely, existentially, freaked.

This is heartbreak.

Now I would be facing an uncertain future without the partner I'd had since I was 18 years old. I'd have to figure out somehow who I was without him and if I could be anyone even worthy of that effort. There was no blueprint. None of my close friends was divorced. I felt, in so many new ways, alone.

This, also, is heartbreak.

Physically, I felt like my body had been plugged into a faulty electrical socket.

In addition to the weight loss, I'd stopped sleeping. I was getting sick: my pancreas wasn't working right. It was hard to think straight.

This, too, is heartbreak, and it's what finally propelled me to seek some answers.

Not only did I want to figure out what was happening to my body, I needed to roll myself upright and get better.

I TRIED TO find refuge everywhere I could, and that meant sometimes in the rational arms of fields like neurogenomics and the psychology of social rejection. The latter sounds like middle school snubs but actually describes a vast ocean of pain that stretches from the playground to broken marriages like my own. The research was both fascinating and unnerving. Despite having lost my mother to cancer when I was in my 20s, I'd never experienced the disorienting sorrow, shame, and peril of losing my life partner.

Romantic heartbreak can cause complex emotional trauma. As with loss associated with the death of a partner, it can dismantle your identity. But the grief of this type of heartbreak is compounded by rejection, which, I came to learn, we humans feel as a deeply evolved threat to our survival.

Breakups are common, and heartbreak is nearly universal. And yet, it wallops us. About 39 percent of all first marriages in the US end in divorce. Psychologists rank this event—or what one 1852 medical textbook called "the slow tortures of connubial disturbance"—as one of the top stressful and consequential life experiences we have, just below the death of a loved one.

For as long as there has been literature, writers have rendered heartbreak—connubial or otherwise—as akin to physical pain and, specifically, to a kind of pain bound to the expectation of more pain. Catullus used the word *excrucior*, the particular and agonizing feeling of being nailed up by your palms, exposed. Susan Sontag, dumped by feckless Jasper Johns when he left a New Year's Eve party with another woman, was similarly graphic: "It hurts to love. It's like giving yourself to be flayed and knowing that at any moment the other person may just walk off with your skin." And here's poet Anne Carson, whose lover of five years told her he no longer felt what he called spin: "Woman caught in a cage of thorns. . . . / unable to stand upright."

There was plenty of heartbreak art—*so much art*—but I wanted science. If this was such a common and devastating experience, why wasn't there a validated protocol for recovery beyond weep-dancing while belting out Gloria Gaynor? Where was the research and what did it say? You'd think after a million years of hominins sighing at the moon over lost love, we would have figured this out by now. Why did nature design us to be so deeply, even operatically, sad? Why was heartbreak so hard to get over? If I learned the answers, maybe I could speed it up and feel better.

MUCH HAS BEEN written about the science of falling in love, but very little about what happens on the other side. Only in recent years has science begun to excavate some of the literal biological pathways of this brand of pain. If you place someone who has recently suffered heartbreak in a scanner, parts of the brain light up that are very closely related to the parts that fire after receiving a burn or an electrical shock. And if some study subjects are unfortunate enough to receive an electric shock after heartbreak, the pain will hurt more. If, however, you shock someone who remains in a loving relationship, and they hold their loved one's hand or gaze at a picture of their devoted squeeze, the pain will hurt much less.

Even more remarkable is the understanding that heartbreak reaches far beyond emotional anguish to influence physical health. People who have suffered lost love face an elevated risk of serious medical woes. It's not just their metaphorically sundered hearts, although cardiac risk is a part of it. Their cells look different; their immune systems falter; even their language skills drop off. Why would evolution equip us with an operating system so easily weakened by an event as common as the denial of love?

"Heartbreak is one of the hidden land mines of human existence," one genomics researcher told me. The language of heartbreak may sometimes sound mundane, but the havoc it inflicts on our brains and bodies is trenchant, profound, and, until recently, understudied. Among the documented downstream effects of rejection, grief, and loneliness are fragmented sleep and fatigue, increased anxiety, poor impulse control, depression, cognitive decline, altered gene expression, and early death.

Why should your immune system care if you've been dumped? What skin does a white blood cell have in the game of love? Plenty, it turns out. Some of our cells listen for loneliness. They adjust their work accordingly, sometimes with devastating consequences for the rest of your potentially feeble, truncated life.

I was eager to reverse course, since my cells, unfortunately, appeared to be listening. New advances in genomics and experimental psychology promised to show me exactly how they were responding and what I might do about it.

I set out to experiment on myself, to see if I could understand the way heartbreak changes our neurons, our bodies, and our sense of ourselves. I would have my nervous system monitored while viewing pictures of my ex. At different points after splitsville, I would measure my threat-mediated biomarkers of inflammation. By better understanding the ailment, I would perhaps find some remedy.

This book traces the general trajectory of heartbreak, from the moment of shock to feelings of rejection, to grief and loneliness, and, finally, toward a measure of repair. All paths through heartbreak are different. Mine was messy and strange and often unexpected. I came to dismiss some of the conventional approaches to recovering from rupture, especially the ideas that you shouldn't form other attachments too quickly and that the key to healing is the commonly traded bromide of "loving yourself first." Both of these exhortations fell short for me. I had to improvise a different course. Ultimately, I would orient toward the far shore of heartbreak through balms much less obvious but supported by evidence: beauty, agency, and purpose.

I would try a regimen of solutions and substances, mostly legal but not entirely. I would travel across the US and to England and Croatia to meet the researchers, practitioners, and ordinary people who were ahead of me in learning how to move on. Some of them were dealing with loss and serious emotional trauma. I'm not equating my pain with theirs, but I believe that we can learn from people in extremis about the systems governing our emotions and our health, and the lessons and potential treatments that apply to all of us. What I found was extraordinary, surprising, and immensely helpful. It would change the way I think about the world, our health, our relationships, and what it means to be human.

I would hike my heart back to life whenever I got a chance. I would try trauma cures, nature cures, psychedelics in a makeshift clinic, companionship of many flavors. *If you don't know knots, make lots*. And before too long I'd end up here, at river mile 119, silt working its way under every fingernail. I knew what wreckage lay upstream. Everything else was a question mark. It was clear there would be no fast heartbreak hack. Also, I hated scorpions. But at least I knew what I needed. For now, it was to lighten my load.

PART ONE

SHATTER

1

BRIDGE TO NOWHERE

The boundary to intimacy is asserted by industry.
—LENA ANDERSSON, *ACTS OF INFIDELITY*

Our love affair began in a hurricane. I'd been in college exactly five weeks. I felt lucky to be there, on Yale's gothic campus surrounded by crisp mid-Atlantic air. I had been ready to leave home. Here, there was so much easy abundance. In the dining hall, a place the size of a hockey stadium, mountains of vegetables rose from the buffet and I was inexplicably dazzled and slaked by the constant stream of cool, amber apple juice. I'd already met friends I loved, admired, and envied. They were like me but they had better home addresses and the easygoing affability that often grows in those zip codes.

Life was busy in the best way. After reading David Halberstam's *The Amateurs* over the summer—a book that was, like all of his books, about men—I joined Yale's freshman women's crew. It was a small group of rowers, including my new best friend, whom I had talked into joining with me. Every afternoon we jogged with our tall and tan teammates to the boathouse. We learned to carry the eight-person shells over our heads and deposit them carefully on the river in a graceful synchronous swoop. We climbed in and grasped our solid, heavy oars. Then we reached with our hands beyond our bent knees and dipped our blades in the smooth water like spoons poised to stir tea. In unison we contracted our cores, released the springs in our legs, and pulled with all our might. The narrow boat glided like a skate blade, quiet and powerful.

I had never felt so much a part of a group, and I liked it. Then there was the water, so much water. On Saturday mornings I jumped on my rusting 10-speed. I rode fast to the harbor on Long Island Sound, where I'd found a work-study job on a schooner that took city kids out to feel the brief freedom of a sea-scented breeze and to learn some biology. At the time, I didn't think about the boats that were to anchor so many threads of my life. I just liked them. We sent small buckets from the schooner to the basement of the Sound to gather pungent benthic layers of mud. We talked about organic pollution causing blooms of algae that suck up oxygen in the water. When this happens, other species can't get enough oxygen and die. I knew I was living life in the oxygen zone.

MY HOME LIFE back in New York City had been much quieter, shared with my mother and a small Lhasa Apso named Albert. We liked to say it backward, Trebla. It sounded like Trouble, and that fit.

But it was my father who delivered me to southern New England that fall in his '70s Dodge van. It was a vehicle better suited to carrying canoes in the mountains than to parallel parking. We'd taken it out west nearly every summer, running rivers by day, camping or sleeping in the van by night. In New Haven, we relayed my boxes of sweaters and cassette tapes up four flights of stone stairs on the freshman quad. I kissed Dad's bearded cheek, grabbed my backpack and hiking boots, and walked downstairs to join a group headed out on a three-day orientation camping trip. I worried about a boy from Jordan wearing a crisp trench coat and dress shoes. I met a tall woman named Ann who wore the same model of hiking boots as me, and they looked like they were the same big size. We would become fast friends. And I met the group's leader, a senior wearing John Lennon glasses and a blue bandanna ringing his head like a line of latitude.

That day, my first day of college, standing in the quad under the bright sun, he and I joked that we had the same last name.

"We must be listed next to each other in the college directory!" he said.

"We should get married some day!" I responded, and we laughed. I probably wouldn't have made that joke if I didn't on some level already feel it could be possible. He was beautiful but approachable: tall and lean, with ropy rock-climber muscles under his shorts and T-shirt. Under the bandanna flowed longish, messy hair, and below that a smile that didn't quit.

This was the first year of the hiking orientation program, which he had started, raising the money himself, because he believed nature was the best place to make new friends. So many freshmen wanted to sign up that there had been a selective lottery. I found out later I hadn't made the cut, but he had rescued my application and put it in the Go pile because I had camping and river-running experience. He, it turned out, liked boats too.

On the bus ride to the Catskills, I sat a couple of rows in front of him and listened to his funny, goofy misadventures about parking his best friend's car on a steep hill and having to winch it out of a forested gully in the middle of the night, about hitchhiking through Arkansas, about fasting in the wilderness and having visions only of pizza. He was both man and boy, worldly and funny and naive. He was someone who could protect a girl in the woods and make her laugh while he was doing it. A boy who was both safe and reckless was, to me, irresistible.

IN 1974, TWO psychologists ran an experiment in Canada that became known as the Creaking Bridge study. Wanting to know which chemicals and impulses in the brain were linked to romantic love, they selected two walking bridges relatively close together: one wide and solid with firm handrails crossing a small rivulet; the other narrow, swaying, and rickety (with sketchy low cables to hold) above swift rapids. They asked a comely research assistant to stand in the middle of one bridge and then the other. Each time, she asked passing men to fill out a questionnaire, offering that each could call her at home if he wanted to talk further (in 1974, it was assumed that most dudes alone on a bridge were straight). Only 2 out of 16 men who crossed the safe bridge called her. But 9 out of 18 men from the rickety bridge followed up.

Since then, numerous studies have also suggested that excitement, novelty, and even anxiety can enhance sexual attraction and romantic love.

Perhaps this is why, five weeks later, I didn't think twice about heading out into the hurricane. The storm was named Gloria. It made me think of dancing to Laura Branigan in middle school. *Will you meet him on the main line / or will you catch him on the rebound?* He'd hatched a plan to climb to the top of the stone bell tower to experience the storm. Mere mortals like us were not allowed near the bells, which resounded every hour throughout campus. But he had a friend named Miles who knew how to pick locks. This thrilled me. It all thrilled me. We packed our sleeping bags and some rope and met Miles after dark outside the tower's thick beveled wooden door. Miles did his thing and the door creaked open. We climbed 284 stone stairs, past the bells to a narrow ladder leading upward. We heaved ourselves through a small portal to an open-air balcony surrounded by elaborately carved finials. For a moment I pictured Kim Novak falling to her death in *Vertigo*. It was a dark night and very windy. It smelled like wet stone.

Wearing rain gear and hiking boots, Miles was wholesome for such a master criminal. His story was remarkable. As I recall it, he grew up poor but mechanically inclined in New Jersey. He learned how to jump ignitions. He stole cars, but returned them. His senior year of high school, he took the SATs and aced them. But his grades weren't great, so he faked a transcript and letters of recommendation, and was admitted to this venerable institution, where he majored and excelled in physics. The outlaw, though, hadn't quite left him.

I was pleased that my crush knew people who were rogues yet had both practical life skills and high test scores. They knew how to pay bills and tie bowlines and trespass into a hurricane. The storm was strong, but not as dramatic as we'd hoped. We watched the sky swirl and the big raindrops turn sideways and then we climbed into our sleeping bags. Mr. Rock Climber tied a rope around his waist and around mine.

"You will not blow away," he said, wrapping a Gore-Texed arm around me. We slept on the narrow ledge in a figure eight of ropes. We were becoming entwined.

It was our creaking bridge moment.

FOR MOST OF our many years of marriage, being closely bound through a life of shared adventures and large and small triumphs and challenges felt like a lovely miracle.

My parents had six marriages between them. His had two. After husbands Number One and Two, and during most of my childhood, my mother conducted an on-again, off-again 15-year affair with a married colleague. I knew of it from about the time I was nine. We used to pass his wife on the street, and my mother would point her out to me and whisper a conspiratorial comment about her hat or her coat and we would titter. True to form, Mom pronounced both her lover's name and his wife's backward, as if that conferred discretion. He was Notlim. She was Erdried.

By comparison, my romantic life starting freshman year was undramatic, well behaved, and stable. It felt to me like living in a sweet foreign land with a much simpler syntax. We moved through college and graduate school, getting engaged on a snowy trail when I was 24, setting up house out in the Rocky Mountains and skiing and kayaking and working. I knew I was fortunate. We lived as a team, feeling lucky and healthy and strong. Our work was good, really good, for both of us. I wrote articles and books about science and nature and other things that made me curious. He worked to protect large natural landscapes and together we played in them. In those early years, before kids, our play was fun and challenging and always drawn against spectacular scenery.

We used to kayak a particular stretch of the Gunnison River in western Colorado. It features several big, technical rapids that scared me, especially one called Boulder Garden. One weekend, the water was high and I didn't like the way the forceful current piled into a giant rock on river left. I could picture my kayak (with me in it, maybe upside down) getting pinned against it. We scouted the rapid and talked through all the moves.

"Do you want me to stand on the rock just in case?" asked the husband.

"Yes!"

In my memory, I did end up too close to that rock, and he did push my boat off it. Or maybe I just imagined it, but I knew he would have. That's what marriage supplied, among other things: a certain amount of Bubble Wrap.

Once our kids arrived, after nine years of marriage, the balance of work and play shifted dramatically to work and parenting. Somehow our tasks grew disappointingly, dismayingly uneven, with more work for him and more of the parenting load for me. Outdoor adventures had been our glue, and now, while still fun, the adventures were fewer and packed with Cheerios, baby carriers, and constant vigilance (mostly mine) for signs of toddler meltdown. We didn't know how to connect so well as a couple in this new reality. Mostly, though, life was busy and rich and meaningful. The husband was a good man, if distracted.

As the years went by, instead of keeping each other company through the rapids and the ski chutes, we were more likely to take turns or go alone. If we were together, he was more likely to say, "I'll see you at the bottom." I thought that must be a measure of healthy self-reliance, or growing up. What I thought then was, I can do this! What I see now is that it was the beginning of the end.

———

THE LOVE THAT began in a sudden burst of weather ended much more slowly. We paid less attention to each other. He would leave for a business trip without saying goodbye, and then he wouldn't call from the road. Even in town, he wasn't great about returning texts and emails. I felt like he didn't help out enough in ways that were important to me. His eyes would light up when younger, prettier women came around. I wasn't always there for him either. I had misgivings about his demanding new job that moved us from Colorado, which I loved, to Washington, DC, which I didn't. I sometimes found it hard to muster the enthusiasm to hear about his workday or his latest athletic training regimen. I had more than a full plate of my own.

In our last decade together, we tried therapy briefly several times. But we were still not feeling supported by each other. When we are depleted, marriage can become a constant tally of each other's inadequacies. As novelist Tessa Hadley puts it, "Marriage simply meant that you hung on to each other through the succession of metamorphoses. Or failed to." We were failing. Somewhere along the line, we had let each other go. I felt sure we could put all the pieces back, but he was already boxing up the puzzle.

The details of what finally drove us apart don't really matter. We were both culpable. I still loved him but I had stopped putting him first. He wanted to find someone else who could. One day, I was cooking dinner before friends arrived, greens draining in the colander, when I asked for an update about a sick relative. He handed me his phone to read an email from his brother. Only a different email came up, one from him to a woman. I had to read it several times before I understood that he was gushingly in love with her. The doorbell rang. It was a while before we could talk about it. Nothing had happened, he said. The email was just a draft. He was confused.

That night, the bottom fell out of my solar plexus, and it would remain somewhere south of normal for a long, long time. He still loved me, he said at first. We stayed together for two more hard years. I still wasn't ready to give us up. By the morning my husband rollered out the back door for good, a small part of me wanted him to go, but a much bigger part of me didn't. My heart was still in it, which is how a heart comes to be broken. Our hearts had been beating together, side by side, through my entire adulthood, and then they weren't.

2

THE HEART

It is a strange thing to feel illness and grief in the
same organ. There is no telling one from the other.
—MARILYNNE ROBINSON, *GILEAD*

My husband moved several long blocks away. For a couple of months, he and I switched off staying with the kids in our marital house, until they left for camp for much of the summer. He readied his new place, after which we'd each have the kids for seven days at a time. Writing this plan had been the worst experience of my life. We worked with a counselor who specializes in this misery. She had also helped us rehearse how to break the splitting-up news to the children, who had no idea this was coming. I didn't want to be apart from them for half the remainder of their childhood. I didn't want to miss half the holidays, half the birthdays, half the vacations, half the weekends, half the dinners. I didn't want any of this. I felt like I was being dragged underwater. I cried through most of our sessions. I railed through others.

"Do people ever get back together because this is so hard?" I asked one day.

"Yes," said the counselor. "It happens. Do you think that's a possibility?"

Both of us swiveled to look at him.

He looked away.

THE HEART CAN break in a lot of ways. It took a while for the cracks to form in mine. But sometimes loss happens all at once, so suddenly that the heart literally stops working. Those initial moments of shock are enough to ignite a chain reaction that can have long-lasting effects.

I'd never been that interested in other people's heartbreaks. But after it happened to me, I couldn't get enough. I turned to literature, to science, to good music, to bad music, and to a wide circle of acquaintances, and then to a wider one. I wanted company. I wanted perspective. I wanted to understand why it hurt so much, this moment of shatter, regardless of whether it happened slowly or fast. It didn't take long before my sense of curiosity started to compete with the pain, if not to assuage it. The heart was an astonishing thing. I was surprised to learn how much the actual beating organ—the seat of legions of metaphors—is implicated in human emotions. It was the beginning of my learning about the comingled destinies of our cells and our passions.

To understand where heartbreak begins in the early moments of shock, we need to start with the remarkable knot of muscle—11 ounces of pure viscera and imagination—that lies at the center of it all.

"It is the world's most banal story. You lose a guy," said Emma, a friend of a friend. In her case, though, banality turned to terrifying physical reality. One day, not long after her boyfriend dumped her, she keeled over from heart failure. She was 41 years old. She survived it. Although her story was still painfully fresh, she was willing to share it.

Emma greeted me at the door of a large house in a posh neighborhood in DC. She worked for the owners and often house-sat while they
were away. By way of introduction, she knew a little bit about my own
heartbreak (Marriage of 25 years! Suddenly alone at age 50!) and welcomed me graciously and with a look of solemn solidarity. Emma had
the place to herself and had laid out a beautiful array of nuts and cheese
and highly textured crackers. She is strikingly beautiful, tall and blonde
and fine boned. Wearing yoga pants and a black off-the-shoulder shirt,
she moved quickly and in bursts, unleashing a coiled-up, skittish energy.
We carried the big tray of food and our glasses of wine into the den and
settled cross-legged into well-upholstered chairs. She employed the large
gestures and precise diction of a trained actress and opera singer, which
she is. Her expressive face shifted from animated to haggard, young to
old, in an instant. Her hair was wild, her eyes wide.

In this big house with her wayward hair, she reminded me of a gen-
tler and more glamorous Miss Havisham, the famous Dickens character who was jilted at the altar and spent the rest of her days wearing her
decaying wedding gown in her gothic manse and generally terrifying
the neighborhood.

I could tell Emma was nervous. In the way common to some thin
women, she barely touched the food. I couldn't stop eating delicate strips
of herring on the difficult, crumbling crackers.

"It was a year ago August," she began. "A storm was coming." She
unspooled her tale over many hours in a nonlinear way. Sometimes
she would spring up like a steenbok to fetch more wine. Eventually she
brought in the bottle, then another. Afternoon turned to evening to
night. We moved into the kitchen to heat up some vegetable soup she'd
made earlier.

"I've always been very sensitive to barometric shifts," she said, her
hands fluttering. "The doctors later told me that day was a perfect storm.
Three threads, coming together. Quite literally, one of the elements was
a storm."

She was 37 when she met the dude, whom I'll call Compeyson, for Miss Havisham's swindler. He was dashing and flawed, which made her feel that he was understanding of her flaws, too. It made her love him more. She opened up to him in a way she hadn't with anyone. They both wanted children. "I wanted to sing to babies and I used to write children's books and I wanted to pass myths and legends on to my kids." Compeyson—I'll just call him C—often consulted in Africa, and they talked of leaving the city to raise a child on the acacia grasslands.

So many plans they hatched. She would wind down her work projects here and they would marry and get pregnant. C wasn't always easy to be with, but Emma loved him and was devoted to their shared future. She worked tirelessly for a couple of years, downsizing her consulting practice so they could leave the country—saving money, not taking time off. By the time she was 40, she was exhausted by the preparations. C told her he needed some space, and took off on a three-month-long trip. They'd been together for two and a half years. When he returned, he said he wanted to date other people. Then she found out he already was. Not long after, he texted her that he'd just gotten a woman pregnant. The baby would be due in August.

They broke up. She drove by his house nearly every day. She would come home and cry and wail and sing opera and the blues to herself. She could really nail Billie Holiday. Emma sang for me, showing how she nailed it. *Good morning, heartache.*

On that fateful August evening, she picked up some take-out food. While driving back to her employers' empty house, she saw a for-sale sign in front of C's place. A moving truck was parked at the curb, likely holding one of her chairs that she'd never retrieved. "I was like, wow, I bet my dishes are in there. And my duvet that she's been sleeping on . . ." Emma was twirling her hair now as her voice trailed off and then came back, throaty.

"I'd given up my clients, my possessions, I had no home, no baby." She was sure she understood the picture fully: the new family was going to be moving to West Africa. The evening storm was building, bringing with it the kind of barometric change that sometimes triggers labor. "I remember thinking, it feels like the kind of night when the moon is full and, oh my God, this baby is coming tonight! I felt it in my bones, the baby that should be named this and the baby that should have been mine."

Somehow, Emma got back to the house and warmed up the takeout, thinking about the storm and the baby and the labor. The labor likely happening at that very moment. It was 8:54 p.m. Then her entire body—Emma's body—contracted. She didn't know if it was a strange empathy thing, a hysterical labor of her own, or her heart finally collapsing under the burden of so much. "My muscles just seized," she said, as if she still couldn't quite believe it happened. "It's so embarrassing!"

In the telling, she was gasping a bit. "There was a moment of stillness where nothing moved. There was so slow a beat in my heart and then it just stopped. It was like my body was cooked, done, folded in on itself. I felt spasms of pain."

She remembered thinking, *Get to the ground*. She crumpled to the floor of the front room. She stared at the ceiling. She couldn't move. "I was fully gripped," she recounted. "I knew then I was having a heart attack. I definitely was lucid." She remembers trying to breathe. "I even actually thought, oh my God, I am doing Lamaze breathing and *she* is probably doing Lamaze breathing too." Emma didn't know where her phone was and she was too ill and ashamed to look for it. Eventually, she crawled very slowly upstairs, thinking she didn't want anyone to find her dead in the foyer.

Our mutual friend, who also worked for the family, found her the next morning, and made her go to the hospital, where the doctor castigated her for not coming in earlier. She was lucky. She was young, and otherwise healthy. Her heart is now mostly working.

Fifteen months later, she was still trying to figure out how to get over C. She was spending a lot of time in nature, kayaking, walking, dancing, singing. But she still looked like a scared animal. It was a look I recognized in myself.

———

EMMA'S DOCTOR BELIEVED she had suffered a classic case of something called Takotsubo cardiomyopathy. I'd been reading up on it for a while. Sometimes called "broken-heart syndrome," it's brought on by sudden distress in otherwise healthy people. Although it's been long known that emotional shock can cause heart failure, it wasn't until 1990 that Japanese researchers were able to use new imaging technology to reveal an unexpected signature of illness, and it wasn't until 2006 that the American Heart Association formally recognized the condition.

In Takotsubo, patients appear to be having a regular heart attack, presenting with chest pain, fluid in the lungs, shortness of breath, and compromised heart function. What makes these attacks different is that the arteries are free from the typical blockages that cause cardiac failure. Instead, a portion of the left ventricle—the heart's main pumping chamber—wildly underperforms, causing it to balloon in compensation. No longer a neat fist, the heart now sprouts a weird distension like an overeager blister. Doctors named the condition Takotsubo after the Japanese lobster trap, which has a narrow neck and bulbous head. In this kind of cardiac event, the heart cells don't necessarily die, but they give up for a while. Approximately 5 percent of patients don't survive the initial event.

Takotsubo, which appears to be caused by a sudden rush of stress hormones stunning the heart, represents about 2 to 7 percent of all sudden-cardiac hospital admissions, but the actual incidence may be higher (diagnostic images aren't always taken). While most patients recover within a couple of months, about 20 percent will suffer complications, including heart failure, arrhythmias, scarring of the heart muscles, and premature death.

Like grief and shock, the condition can be shared, reflecting the many reasons the heart breaks. We can experience the collective grief wrought by war, violence, and natural disasters, the grief of losing species and habitats to which we feel deeply akin; there's the collective and individual grief of racial and social injustice, the individual grief of core relationships blown up through death or chronic illness or by someone's choice.

After a major earthquake in the Niigata Prefecture in Japan in 2004, researchers found a 24-fold increase in Takotsubo cases within four weeks. In 2011, researchers from the University of Arkansas examined 22,000 cases and found spikes in Vermont, which had suffered lethal, unexpected flooding, and Missouri after the Joplin tornado. And an Ohio study found a 4-fold increase in cases during the Covid-19 pandemic (the patients did not have Covid).

The case literature of Takotsubo patients includes recent widows, women whose children or pets just died, and people undergoing other extreme stressors. One woman reportedly suffered Takotsubo after accidentally eating a huge amount of wasabi. One 56-year-old man was found to suffer from Takotsubo after his favorite soccer team lost the 2012 European Cup in Poland. He made a full recovery. After Chile flubbed the last penalty kick in a championship match against Brazil in 2014, a 58-year-old man went into classic cardiac arrest, in which something, typically high blood pressure brought on by emotional or physical stress, causes a piece of plaque to break loose and block a major artery. He was rushed to the hospital, where he died. A little over an hour later, his 64-year-old wife collapsed with chest pain, tightness, shortness of breath, and heart failure; she was experiencing Takotsubo. She survived. (As Bill Shankly, the former manager of the Liverpool Football Club, once put it, "Some people believe football is a matter of life and death. . . . I can assure you it is much, much more important than that.")

Science bears out the close relationship between heart health and love. Karl Pearson, the British founder of biostatistics, examined gravestones in the early 1900s, noting that husbands and wives often died within a year of each other. More recent research shows that people unhappy in love—either in bad relationships or in broken ones—suffer higher rates of heart disease. A survey of 43 million medical records in Denmark found that in the year following a romantic breakup, men between the ages of 30 and 65 experience a 25 percent increased risk of a heart attack, and women experience a 45 percent higher risk. In both sexes, the risk remains elevated by 9 to 24 percent even nine years later. Love protects your heart, while loss weakens it, sometimes forever. As physician Sir William Osler put it in 1908: "The tragedies of life are largely arterial."

With all the grief going around these days—whether from foiled love or global contagion or social injustice or the climate or a host of other things—it's a wonder more of us don't collapse from the emotional toll. Why don't we? There are many reasons to love estrogen. One of them is that it protects us from sudden shock, which is why 80 percent of Takotsubo cases occur in postmenopausal women. Perhaps evolution designed it that way so women's hearts wouldn't seize up during our highest and scariest exertion event, giving birth, when our estrogen levels are through the roof. But after the childbearing years, good luck.

Martin Luther once said, "Faith resides under the left nipple." Takotsubo is the metaphor of a broken heart made real. It renders obvious a truth that is more subtle: our bodies want us to feel safe and to feel loved. What happens to us when we lose that attachment is a central theme of this book.

I WANTED TO learn more about what had happened in Joplin, Missouri, on May 22, 2011, when an EF-5 tornado ripped through town. As someone interested in the therapeutic benefits of nature, I was especially curious about the butterfly garden created to help the community heal in the wake of loss and grief. And I'd also heard about the uptick of Takotsubo cases. The wrath of nature was causing literal heartbreak, which was in turn salved by nature. I reached out to a man who became caregiver to both wounded trees and wounded people.

On that fateful day, Chris Cotten had been on the job as the city's director of Parks and Recreation for just 86 days. In the afternoon, he visited Lowe's, picking up fencing material for his family's new yard. Just west of him near the Kansas border, a bulwark of thunderstorms traveling northward from the Gulf of Mexico had crashed into a cool mass of air traveling south from the jet stream. By 4:12 p.m., three churning supercell thunderstorms were generating strong wind shear as warm layers of air rolled over cooler winds near the ground. A rotating mesocyclone tipped into vertical position as it continued on its inexorable path eastward toward town.

Cotten's wife and four kids were safe outside of town. He made it back to his house and watched the sky churn yellow as huge balls of hail dropped onto his truck. He heard the sirens. The air smelled funny, of chemicals and underlayers of soil. It was the smell of destruction. By 5:39, anyone who was still outside could see the three-quarter-mile-wide twister that had formed into a perfect, horrifying, prairie-eating comma.

He waited until it grew quiet, then he headed out to inspect the tree damage to the parks under his care. But he couldn't even find the parks. Instead, he saw entire neighborhoods flattened, roads cluttered and unrecognizable, debris everywhere, and people shrieking. An EF-5 tornado is as bad as it gets. It is also incredibly rare, making up about one-tenth of 1 percent of all tornados. The storm had taken its time ripping up Range Line Road, a major thoroughfare, taking a number of stores and businesses with it. At the Home Depot, the roof had popped off like a soda can tab, knocking over the store's walls like falling dominoes, killing seven people. The tornado lifted St. John's Regional Medical Center four inches off its foundation and knocked out all the complex's power systems, killing a half dozen people on life support.

As a city employee and former lifeguard with first-aid experience, Cotten gathered another park worker and together they converted a public hall into a triage hospital. For the next 48 hours, Cotten carried in bloody bodies, laying them in rows on the stage. He sorted out the injured for medevacs, then assisted surgeries on banquet tables, then ran out to perform search and rescue. Zipping the body of an older man into a bag, he watched parts of the man's brain fall out. He pulled roofing tacks out of a kid's back and listened to a bloodied girl screaming *Please stop* over and over again. The traveling Piccadilly Circus was in town. At one point, it offered up two adult elephants to help remove heavy debris from the roadways to clear a path for first responders and the arriving National Guard.

Eventually, the death toll would reach 161, the worst in the US since a 1953 tornado tore through Flint, Michigan. Cotten did finally assess the tree damage: 18,000 gone. Some of those still standing were completely debarked.

Many of the survivors testified to their faith in God, the compassion of neighbors, the trauma, the exhaustion, the grief. But others spoke of something more surprising: the presence of butterflies, spectral and protective. A little girl recounted how her father had lain across her in the yard grasping at the sod. His shoes were pulled from his feet. When the storm lifted, leaving them unharmed, she said, "Daddy, it was okay, a butterfly was holding us down." Another girl who was lifted into the air described how a butterfly wrapped its wings around her and brought her back to earth. A child who emerged unscratched from a car filled with glass and debris described a similar experience in recorded accounts.

Chris Cotten knew what to do. He took a park and residential street at the epicenter of the devastation and turned them into a place of healing: a butterfly garden. Today, the site features black steel posts and beams that create 3D outlines of the houses that once stood there. Sedges and small shrubs provide habitat for prairie butterflies, and a journal sits attached by a chain to a bench where people can sit, reflect, and write. And of course, there are the trees, thousands of them, newly replanted. Cotten's team received input not only from the community but from the emerging field of "resilience studies." As someone looking for lessons of resilience, I was eager to learn more.

Keith Tidball, a researcher and fellow at Cornell's Atkinson Center for a Sustainable Future, who worked on the project, told me he is interested in two kinds of resilience: ecological and psychological. Often, he says, the two are intertwined. A natural disaster, or a human-caused one, like a war zone, leaves both people and landscapes ravaged. It's not uncommon for people in affected communities to suffer from post-traumatic stress, grief, depression, anxiety, rumination, and suicidal thoughts. In the worst disasters, up to 30 percent of the local population may show some of these symptoms for a period of time, with children being the most vulnerable. Tidball studied the effects of tree planting on trauma- tized residents in New Orleans after Hurricane Katrina. He found the saplings became symbols of ecological and social renewal, a response to the narrative of New Orleans as a failed, broken city. The civilian acts of planting served as gathering opportunities, and the trees themselves became icons of resurrection.

"The loss of trees is a major heartbreak and disruptor," he said, "so the act of stewarding them back is fundamental to recovery." In a crisis, nature reminds us that we are not the center of everything, and also that we are all connected.

Often people don't even realize how much they love and need the trees until they are gone. Tidball calls this yearning "urgent biophilia," and it was a phrase that kept coming back to me. We need the bosom of the natural world, and sometimes we need it now, and in every town and in every neighborhood. After heartbreak—whether over a devastated landscape or a personal loss or a global crisis like a pandemic—it is often nature to which we turn. The more uprooted we feel, the more we need the literal rootedness of things with roots.

"When you hurt, nature heals, and that's what this is about," said Tidball.

AT THE FIVE-YEAR anniversary of the tornado, hundreds of people turned out at the Butterfly Garden & Overlook for a children's fun run, a silent hike, a community picnic, and commemorative talks. Chris Cotten was not one of them. He moved to Kansas after the park was dedicated in 2014, and he can't bring himself to go back.

"I just couldn't deal with the tornado anymore," said Cotten, who, like many first responders there, has undergone treatment for recurrent nightmares, anger issues, and anxiety. Last year, his marriage unraveled, more collateral damage from the tornado, he said. He had become emotionally withdrawn and distant. Rebuilding the parks took over his life; he wasn't there for his family, he said.

Perhaps because he worked so hard to make a healing space for other people, the garden just didn't bring him peace. "I take comfort that it helps other people," he said. "But I don't know if I'll ever take comfort or get over what happened." As bad as the storm itself was, he said the trauma of his divorce was just as bad.

"Really?" I asked him.

"Yes," he said, explaining that both experiences clashed and magnified, much like the storm cells on that May day in 2011.

It was a sobering thought that made my stomach sink in both dread and recognition. I didn't have to live through a literal tornado, but it was clear that for many of us, heartbreak tears up all familiar terrain, leaving behind trauma, physiological chaos, and a shattered identity.

Nature appeared to be helping Emma recover, but it wasn't enough to help Chris Cotten.

Would it help me?

I hoped so. I didn't really have a lot of other ideas.

3

HINDU KUSH

Redemption preserves itself in a small crack
in the continuum of catastrophe.
—WALTER BENJAMIN, "CENTRAL PARK"

A month after my husband moved out, we still had one last-gasp, long-planned family vacation in Mexico with his family. I didn't want to go, but the kids wanted me to, and I hated the idea of them being on vacation without me. I agreed to fly down for a few days before heading to Colorado for meetings. It was the in-laws' anniversary, at a resort near the beach on the Mayan coast. I arrived a day after everyone else. Some members of the extended family were warm, some cool. Nobody really knew how to act, including me.

My daughter, at 13, stuck to me like Velcro, or maybe I to her. We walked the paths holding hands, trying out different pools, Jacuzzis, water-yoga classes, and personal-care products. We are both cursed with the same shape-shifting hair. In humid climates, it has to be wrestled like a wild beast. She and I shared a room, with my son and still-husband-in-name next door. I gave him as wide a berth as possible, which actually wasn't all that different from the last many months of a marriage characterized by parallel existences, wariness, and distrust. Now we both drank chalices of margaritas from our respective perches at opposite ends of large family tables.

I didn't want to hang out with him, but I didn't know how to be without him, either. Vacations had always been when things between us felt best, our rhythms in sync, when we set our gaze upon each other—and on our family—and when the feelings of neglect, hassle, and the corrosive inequity of chores receded. We enjoyed doing the same things outside, like towing our game and trusting toddlers in a bicycle trailer for three days across the Idaho panhandle and riding horses deep into the Montana wilderness. I'd been game and trusting, too, and I was finding it disorienting and nonsensical to go from feeling like your partner would save your life on a mountaintop to realizing he wanted to live his life without you.

I married one of three brothers, and I'm close to the other wives. Early one morning before the kids woke, we three sat under a pool pagoda.

"The only advice I have for you is that we are here for you," said Ana. "Don't do it alone. We love you. You are always our sister."

The other sister-in-law, Lisa, is also one of my closest friends. I've known her since I was 19, when she was a young journalist dating a rakish kayaker, and I was working at a kayaking store in Colorado the summer after freshman year. She first met her future husband, my brother-in-law Peter, at my wedding. Many years later, I reintroduced them. I thought they'd get along, and I also thought it would be excellent to have her company at family gatherings. I was right on both counts. Lisa is eight years older than I and much, much wiser. She is part sister, part wilderness buddy, part spiritual adviser. And she probably knows me better than anyone. While I took up with my future husband at 18, she took up with hers like a normal person, well into her 30s. She knew heartbreak. She also understood the pretzels one folds oneself into for love. She had followed one soulful man down perilous class-IV rapids. Still, he cheated on her. She moved into another boyfriend's cabin without running water, where her hay fever was so bad she sneezed her way through sex. But by the end of it all, she knew what she wanted and who she was. One of the things she wanted was to kayak gentle waves with girlfriends like me and then wear pretty sandals. She also likes to speak her mind.

"Here's the thing about you," she said to me by the pool. "What I've often said to Peter is, how come we make all the mistakes and you guys never make any? Overt fantasticness is what you both have been and now that road is gone. You have been in a traditional marriage that went cold and you've gotten really good at feeling numb. I think it's going to be hard to un-numbify," she continued. "And you don't know anything about men. And making mistakes is really good, actually. I hope you mess around with some interesting people."

I nodded but it all seemed impossible. I hadn't known I was numb. I did want to un-numbify, except everything now hurt.

After a morning of ocean swimming, I hugged my kids goodbye, breathing in the sea-salt smell of their damp heads. I left the resort in a chauffeured Suburban proffered by my in-laws like a parting gift to a reject on *The Bachelor*. I wondered if I would ever see them again. It seemed conceivable that I wouldn't, and yet they had been a loving part of my family for three decades. I had been present at nearly every Thanksgiving, every big anniversary, often in a nice hotel or stately house. Now I felt like I was being politely escorted off the property.

From the back seat, I watched the landscape turn from manicured and irrigated to arid, neglected, unfragrant. I had the strongest vision of being turned out of the fantasyland castle. The big oak doors creaked behind me and closed shut. I would now be crossing a thorny wood. I felt about 16 years old, alone, with few wiles and few allies. In one afternoon, I was plummeting back to the precarious middle class, where exactly no one was waiting for me. What would happen to me now? I had absolutely no idea.

I flew to Denver and switched to a little jet bound for the small desert city of Grand Junction, Colorado, where I rented a car for the two-hour drive to the Roaring Fork Valley. It was after midnight. There were few travelers on the road, just scrubby sage and a dark, fathomless sky. High beams occasionally zoomed past me. I had the sensation of hurtling through space upside down. I'd have to dodge the astral mule deer on my own.

I WAS HEADED to Aspen to give a couple of talks as part of a creative-class workshop aimed at public health and sustainability. I'd just written a book about the science behind why being in nature makes us healthier, and the perk was that I sometimes got invited to ridiculously pretty places to discuss it. For several days, I hiked every morning in the shadow of the West Elk Mountains, then panelized with activists, social entrepreneurs, and academics. Not for the first time, I thought, I am lucky for this work.

It was a particularly sweet junket: a few days at a nice hotel on the edge of town, free food, minor famous people doing yoga in the meadows, all ringed by mountains. I'd heard Helen Fisher would be there. She's a biological anthropologist who studies the neurotransmitters of love as a research fellow at the Kinsey Institute and has written several well-regarded books, including *Anatomy of Love: A Natural History of Mating, Marriage, and Why We Stray*. Aware that she is one of the few researchers to study the brains of people who have been dumped in romantic relationships, I immediately made plans to accost her. I didn't yet know I'd be writing a book about heartbreak. But my heart was breaking and it had never done that before. I just knew I was feeling stunned and scared in a way I never had. Maybe she could help me figure out why my brain was exploding.

"Of course I'll talk to you!" she said by email. "Come over!"

I knocked on her door at the hotel, and she invited me in with a gesture that was both elegant and warm. We kicked off our shoes and sat on her couch by the window. I told her a bit of my story and that I was trying to learn what was happening in my brain.

"Oh, Kiddo!" she said. "I'm very sorry. It's very different to be left than to do the leaving." She appeared like a nerd-mom-counselor just when I needed one. My own mother had been dead for 20 years, and I missed her. Fisher was vibrant, well dressed, alluring, 20 years older than me, and, I noticed, unmarried. How did her skin look so great? I pulled out my notebook.

"Are you sleeping?" she asked.

"No."

"Are you eating?"

"Trying. I've lost 20 pounds."

"Okay, I can tell you why."

"Tell me everything."

She poured us each a glass of water. "One of the most painful experiences that a human being can suffer is to lose a life partner," she said. Despite that, she explained, it's been vastly underexamined as a topic of study. Many scientists, she said, simply underestimate the power of heartbreak, but it's also more alluring to track the excitatory state of falling in love. Fisher herself has done plenty of that. You've probably heard about this work, or read the stories that come out every Valentine's Day about our reward networks and the addictive, delicious qualities of infatuation and lust, about how neurotransmitters like dopamine and oxytocin keep us euphoric, transfixed, hyperfocused on our love object, greedy for every touch, willing to take risks.

But after years of tracking the brains of the suckers who fall in love, Fisher thought it would be interesting to see what happened to them once they've tumbled out the other side. She herself has been there, and so have most people, she reminded me.

For a paper published in the *Journal of Neurophysiology* in 2010, Fisher and her colleagues put 15 people who had recently been dumped in a brain scanner. They took images of each subject's brain as the person viewed a photo of their rejecter and as they viewed a photo of a neutral, familiar person. Interestingly, while viewing rejecters, their brains looked similar to those still experiencing intense feelings of romantic love.

"When you've been rejected in love, it doesn't mean that you stop loving the person," she explained. "In fact, when there are barriers to the relationship you can even love the person more. Just because you've been dumped doesn't mean you just stop."

In the study, brain regions lit up that govern cravings and emotional regulation, including the ventral tegmental area (VTA), ventral striatum, and cingulate gyrus. These mid-brain regions, necessary for feeling romantic love (and, Fisher added, for fostering cocaine addiction), sit right next to regions governing thirst and hunger. Love isn't an emotion, she contends; it's a survival drive mediated by our neurochemical reward system. "Intense romantic love and feelings of attachment evolved so that you can send your DNA on into tomorrow," she said. "They're very primordial brain systems and they do not die quickly."

It reminded me of a passage from Rachel Cusk's divorce memoir, *Aftermath*. "Grief is not love but it is like love. This is romance's estranged cousin, a cruel character, all sleeplessness and adrenaline unsweetened by hope."

If love is an addiction, it can be a constructive one, compelling us toward one another. But when love is not returned, destructive emotions and impulses pile on. In addition to finding activity in parts of the brain linked to craving and addiction (as in falling in love), Fisher's team also saw activation in the insular cortex and the anterior cingulate cortex, which process physical pain. These regions also light up when you have a toothache, said Fisher. The difference is that with heartbreak, the pain can last and last.

"If you get dumped on Monday, it could be days, weeks, months, *years* before you really get over the pain," she said. "So it's a pain that is actually stronger because it's also mixed up with all these feelings of anger and rage."

That's an awful lot going on in one brain. It made my eyeballs hurt.

"It's called abandonment rage," she continued.

Yes.

In Fisher's study, she found that the subjects thought about their rejecting beloveds for a full 85 percent of their waking hours. They also acted pretty nuts, reporting "lack of emotion control on a regular basis since the initial breakup, in all cases occurring regularly for weeks or months," the researchers wrote. "This included inappropriate phoning, writing or e-mailing, pleading for reconciliation, sobbing for hours, drinking too much, and/or making dramatic entrances and exits into the rejecter's home, place of work, or social space to express anger, despair or passionate love."

When it comes to heartbreak, many of us become uncharacteristically histrionic.

Sometimes we become suicidal. Among adolescents in the US, breaking up is the largest risk factor for attempted suicide, and among successful suicides in adults, intimate-partner problems are a factor 27 percent of the time, far more than any other, including poor health, financial trouble, and eviction. In India, about 10 people kill themselves every day over lost love.

"I think nature has overdone it," Fisher told me. "Why is it that we struggle so badly? Why can't we just walk off and find a new partner and be happy again? But when you think about it, when you've been dumped, you have generally lost economic power, financial power, perhaps a home. You may have lost children and a dog and a cat. Every moment reminds you of the situation where you used to get up and there was somebody there and you had breakfast together, and now there's nobody there anymore."

None of this was making me feel the slightest bit better. But she was on a roll.

"And by the way, I mean you have not only lost your partner, but you've lost time, you've lost opportunity, you've got to start over again and if you haven't had children you have lost an opportunity to breed."

I said I was grateful at least the breeding part was behind me.

"Well, there's no good time to get dumped," she said. "Ever. No. Older people struggle just as much as younger people. In fact the brain circuitry for romantic love really doesn't change whether you're 18, 48, or 78." She shook her head. "Almost nobody gets out of love alive."

Fisher stretched out her legs and continued. She explained there are two basic neurological stages of getting dumped: protest and resignation. During the protest stage, many people try to win their beloved back. "One tries to negotiate. It's a little bit like a dog who is taken away from its mother and put in the kitchen by itself. What does it do? It hurls itself at the door. It runs around in circles and barks and barks and barks. That pretty much sums it up."

This behavior, she said, is based on a cocktail of extra dopamine and norepinephrine flooding your brain. You're searching for what you're missing, and you're scared. I could relate to the manic part. I felt like I'd been plugged into an amplifier these past months. This was, she explained, hypervigilance in response to one's new threat-filled state. You obsessively go over a new calculus of survival. This helps explain the sleeplessness and the agitation. Your brain is also working to figure out your gains and your losses. "So you lie in bed at night thinking, who gets the dog?" she continued. "Where do I go for Christmas or whatever my holidays are? Who will be my friends and who won't? You're trying to figure out what you can learn from this."

During the resignation stage, she said, people largely give up the protests and the bargaining. This is when the dopamine drops off, and so does serotonin, a neurotransmitter often linked to feelings of well-being.

"Yeah, I'm there," I said, although I wasn't completely convinced about the resignation. "You sound like you're there," she said. "Once you're there, it's lethargy and of course a lot of tears. Now some people will drink too much, drive too fast, or hole up and watch TV. Other people will talk their heads off about it. Neither is very good." She peered at me through bright, caramel, sympathetic eyes. "You have to build a scenario of what happened," she suggested. "Once you've got your narrative together, then you can throw it out. I remember one time, a man dumped me. I never could figure out why. And I realized finally, *Helen, you're never going to know. So make it up.* So I made up a scenario."

All of these steps sounded painful and horrible, but they also seemed like a huge pain in the ass. What a lot of work, all this bargaining and raging and disposable meaning-making. And what, I wondered, about the person who does the leaving? Does he really just skate away, relatively untroubled, as my husband seemed to be doing?

"Nobody knows," said Fisher. "And when I was really studying rejection, I was looking at it through the academic literature. There's almost none about the person who does the dumping."

I'm sure I looked exasperated. It all just seemed unfair. And later I would find out that women seem to feel the emotions of heartbreak more intensely (although men can suffer worse health outcomes; stay tuned). Research suggests that women in heterosexual relationships remember both positive and negative events in much more detail than their male partners; women ruminate or think about their love troubles more than men, and they get more easily depressed. Men and women also express their pain differently, with men more likely to turn violent when they are spurned. Their victims: the women.

As to the weight loss, she explained, it's a combination of anxiety (our metabolism speeds up in preparation for facing peril), forgetting to eat well (see: depression), and stress hormones like cortisol shutting down normal digestion.

"Listen, Kiddo," she said, "the day will come when you'll look back and say you're sorry he didn't leave sooner. You'll get a whole new life and maybe a new man. I mean you're way too young to go without another man."

I loved it that she thought I was young. I made a mental note to hang out with more 70-year-olds. But a new man was the last thing I wanted. I just wanted to sleep.

I'm sure I looked skeptical. "Falling in love is a beautiful feeling!" she said, revealing she was currently in mid-swoon with a new love interest herself. "I will convince you! What are you doing for dinner?"

ON MY WAY to meet Fisher in town a few hours later, I passed a cannabis dispensary and decided to duck in. There were several, unabashedly splayed among the expensive boutiques in Colorado's new ganjaconomy. I'd never been in one before. It looked like a bare-bones drugstore. There were no other customers.

"What have you got for sleep?" I asked a young guy behind the counter. He looked comically like someone who would work in a weed dispensary: scraggly beard, underexercised, laconic. He slid a canister of gummy bears—strain Hindu Kush—across the Formica. "Start with half a gummy," he said. "We take cash."

Fisher looked beautiful in a sweater and scarf, her smooth blond hair neatly cradling her face. The restaurant was tasteful, bright, spare. We sat at an elevated table and ordered white wine and small plates of salads and aubergines. It no longer felt like an interview. While Fisher is willing to talk about heartbreak, what really ignites her is love. She is, after all, the dopamine queen. She happily told me about the new man in her life.

"There's nothing better than a nice big dopamine rush for the brain," she said. "The dopamine system is associated with focus, with motivation, with optimism, with well-being. And, you know, sex is good for people."

If you are a dopamine junkie, naturally this is what you are going to think. But I had my doubts, or at least an instinct to protect my newly flayed heart. Also, I felt too old for this, but saying it to her made me feel ridiculous.

"Romantic love is like a sleeping cat," she said, smiling. "It can be awakened at any time at all. Isn't that beautiful? It's beautiful."

As we finished the wine and paid the bill, she issued a few caveats. Love was grand, but I should be careful out there. "The bottom line is when you have sex with somebody, any stimulation of the genitals drives up the dopamine system and you can fall in love. With orgasm, there's a real flood of oxytocin and vasopressin linked with feelings of attachment, so casual sex is not casual."

Duly noted.

During our long dinner, the town had grown dark. The sprinklers had turned on, pelting the sidewalks with water. When we stepped outside, Fisher reached for my hand and we ran through the spray in our sandals, wobbling, laughing. Her joy in life and love was infectious. *Relax*, she was saying. *Open your heart right back up*. It was almost starting to make me feel better. I was holding hands with Venus, plus I knew I had a gummy bear awaiting me.

As we swanned our way back to the hotel, she looked at me and got serious for a minute. "What you don't want, Kiddo, is to get dumped too soon by another guy."

With that, the goddess of love kissed my cheek and soared to her balcony.

4

A COSTLY LIFE EVENT

Single women have a dreadful
propensity for being poor.
—JANE AUSTEN, LETTER TO HER NIECE, 1817

Several weeks came and went in a swirl of deadlines, divorce mediation sessions, ironing name tags onto socks for camp. I'd brought a stomach bug home from Mexico, and I lost another five pounds. I weighed 105, the lowest since I was about 12. This might work for a supermodel, but it was not a good look for a 50-year-old journalist. I was watching myself turn into the spurned *poverella* from Elena Ferrante's *The Days of Abandonment*: "Now she came down the stairs stiffly, her body withered. She lost the fullness of her bosom, of her hips, of her thighs, she lost her broad jovial face, her bright smile. She became skin over bones, her eyes drowning in violet wells, her hands damp spider webs."

I felt power-washed by sadness and anxiety. I looked like a stray animal who was trying to paw her way out of a kill-shelter. I would walk our husky mix around the neighborhood and burst into tears. The husband and I were making separate plans for the summer. While the kids went to camp, he'd stay in DC and I'd head back to Colorado. I would work from Boulder, house-sitting for a friend, and try to learn how to take deep breaths again. The previous summer during camp, we had been staggering in and out of couples therapy, where I cried and he looked like he'd rather be in a drainage pipe. This time, he didn't ask much about my plans and I didn't ask much about his, even though being apart felt like a vacuum sucking out my insides.

In 25 years of marriage, we'd never been apart for more than a week or two. As I was learning, this unfamiliar separation is enough to alarm one's nervous system, regardless of how well you were or weren't getting along. Studies in monogamous titi monkeys, prairie voles, and humans all show that when cohabiting mates are separated, each individual's stress hormones increase. Even zebra fish sense when their mates are missing. Their lateral tactile sensors register the absence of other fish, triggering release of a stress-related hormone and a fear response. They know their safety depends on being together.

In humans, when one partner bolts—even on just a work trip—there's some indication that the one left behind suffers more distress in the form of worse sleep, higher cortisol levels, and greater feelings of negativity. When a couple live together, their bodies coregulate to the point that their heart rhythms and other biological processes align. One recent study looked at brain wave patterns in male-female long-term partners versus male-female strangers during social interaction. When the couples (but not the strangers) hung out, they produced gamma-wave synchrony in parts of their brains associated with social connection. This was true whether or not they even spoke. Our ability to sync up neurally with people close to us shows how deeply and unconsciously these attachments take root.

My body knew something was up. The kids left for camp. My house was empty. Where were my zebra fish? What had happened to that tall, rangy zebra fish with the stinky hiking boots?

It was late June. Washington was oppressively hot. I couldn't wait to get out of there, to escape my life.

WHEN I ARRIVED in Boulder, where I used to live, I had some explaining to do. Many of my friends hadn't yet heard the news about my marriage, and I had to repeat it over and over, watching the shock hit. "That's impossible," they'd say. "Not you guys." The pep talks I was hoping for were largely lacking. My friends were all married, and they looked as confused and scared as I felt. "What are you going to do?" they'd ask. I realized I probably needed to hang out with more divorced people, models of strength and survival, but I hardly knew any.

Sleep remained elusive. The gummies helped for the first half of the night, but not the second. I still had that weird frenetic energy, that hypervigilance Helen Fisher described, a buzz saw with no wood to cut. My sister-in-law Lisa lives near Boulder. She and Peter do Boulder things like drive a fully electric car and flush the toilet with recycled bathwater. Lisa helped me in every way. I cried in front of her pretty much every day. We sometimes wrote side by side. We exercised together, we ate together, we gabbed. I even made her jog braless with me while I recorded how it felt for a podcast episode I was making about the invention of the sports bra, an engineering feat that literally changed the destiny of women's participation in athletics. "God, my nipples keep changing direction!" she yelled.

I flew to Salt Lake City for a few days to record some interviews for another podcast. Just flying over the Rockies, sensing all the space and oxygen and rushing rivers down there, opened up a little room between my ribs. One of my interviews was with Stacy Bare, who jokes that if you're a guy with a name like a girl, it helps to be burly and six-foot-seven. I've known Stacy for years. He's always fed me good story ideas about people healing in nature. A veteran of the war in Iraq, he credits mountain climbing with saving him from suicide upon his return, and he's dedicated his life ever since to getting more veterans outside, sometimes with electrodes on their heads to advance science. Bearded, lumbering, he embraced me in that big-armed hug of his, and we sat down to talk about healing trauma.

After a while, I revealed that my marriage had come apart. I told him I didn't want to compare my experience in any way with his, but I was noticing a similarity of symptoms to PTSD: the hypervigilance, the sleeplessness, the anxiety. I asked him if he thought divorce could be similar.

He told me he was very sorry to hear it. And yes, "not only is the divorce traumatic, but it's happening in a society that also doesn't know how to deal with trauma, or separation, or moving on from heartbreak. With a divorce, you wake up every morning and that impact is still happening, it's fresh all over again, like when friends die, you have to find ways to integrate it to move on."

I'd learned about nature's benefits for a lot of woes and afflictions in my last book—war trauma, high blood pressure, myopia, brain injuries, attention deficit disorder, cancer, depression, and so on, but I hadn't run into anyone specifically experiencing heartbreak. I asked him about it.

"One hundred percent it can help with heartbreak," he said. "The wilderness is where I've gone for both celebration and commiseration. It can support finding who we are again and building that up. You've got to find your new geographies."

I hugged Stacy the Giant and continued to follow the trail of crumbs.

BY USING THE word *trauma*, Stacy was putting a label on what Helen Fisher had described: the stress hormones, the weight loss, the extreme agitation. It was all linked to feeling unsafe. Cotten had experienced it, and so had many Miss Havishams. And trauma, I was to learn, affects more than the heart. The focus of my trip to Salt Lake City was supposed to be the nature-loving veterans, but when I heard about a professor at the University of Utah who studies the physical and psychological effects of relationships, I made a detour. Bert Uchino would prefer to study the salubrious benefits of good marriages (and he, lucky duck, has been in one for 25 years), but since many core relationships involve negative aspects, he's expanded his purview.

Uchino's office sits in the brutalist concrete psych tower on campus, sometimes called by its occupants the Tower of Doom, but its views of the scenic city and Wasatch foothills make up for it. Although Uchino is a calm and gracious host, I had a feeling this would not be a fun visit for me. First, he told me how great it is to be well married. If you have a partner who revels in your successes and consoles you in your failures, you feel protected and seen and loved. You modulate each other's stress levels. Partners help us maintain a stable self-concept. Through mirroring each other's emotions, empathetic responding, and physical connection, intimate relationships help us remember we are real and okay. There are, said Uchino, clear, tangible benefits to this that manifest in your cells, your arteries, your organs. Scores of robust studies across different cultures have shown that married people live longer, experience fewer cancers, strokes, and heart attacks, are more likely to survive serious illnesses, and are less likely to be depressed and overweight.

A large Danish study followed 21,000 participants with a mean age of 52 for seven years. After adjusting for socioeconomic status, lifestyle factors, and psychiatric problems, researchers found that mortality increased 60 to 70 percent in the most socially isolated people. The strongest factor in the higher death rate: not having a spouse or a partner.

Such observations are not new. In 1858, a British physician named William Farr studied the census tables and concluded: "Marriage is a healthy estate. The single individual is more likely to be wrecked on his voyage than the lives joined in matrimony."

And while it's important to point out that mentally and physically healthy people are more likely to get married and stay married in the first place, the good-genes theory doesn't explain the full effect.

It's not just that married people help each other through good times and bad. There is a less romantic, purely pragmatic reason for the health boost. The single largest determinant of health in the US is wealth. Married people are more likely to accumulate resources, afford good healthcare and insurance, and live in healthful environments. (The income gap is strongest between married and unmarried women because women make so much less.) If these married people are lucky enough to love and feel loved, Uchino said, they report less pain, to boot.

I was like, *Please stop talking.*

But then came slightly more comforting news to my jaundiced heart: not all marriages are so helpful and healthful, because not all marriages are good. In fact, about half of married people say their marriages are either so-so or worse than so-so. Confusingly, it's the so-so people who may face worse health problems. Uchino explained: When a marriage is clearly lousy and your partner withholds affection, you may develop escape or coping or compensating strategies. You know what to expect and what your marriage can and can't deliver. One friend of mine calls her marriage "stable miserable," and she actually manages pretty well. But when your partner is unreliable in offering love and support— Uchino calls this an ambivalent marriage—the pretty road to health appears far more treacherous. You get crushed by disappointment over and over again.

This so-so zone is where my marriage had lately been. Sometimes I felt like my husband was there for me, but often he was galaxies away, leaving me hurt and confused. There were no clear discordant notes, just an absence of full-throated harmony. The writer Nicole Krauss writes of the "boundless loneliness" of this kind of marriage in her novel *Forest Dark*: "Conflict was not allowed between us, let alone fury; everything had to go unspoken, while the surface remained passive." Susan Sontag called marriage "an institution committed to the dulling of feeling." But your body doesn't stay passive or dull. It knows you're in trouble.

One recent study looking at 30 years of data in the US from 19,000 married people through old age found that those in strong unions were the ones driving the marital health boost; they had a 20 percent reduced risk of dying early compared to people in average marriages. Caveat: Virtually all the large-scale studies looking into long-term health effects are of heterosexual couples. But there is evidence that women in same-sex marriages may be happier than women in heterosexual marriages who end up burdened with far more domestic labor than their male partners. Early in the Covid-19 lockdowns, one amusing *New York Times* headline captured a perfect snapshot of modern heterosexual marriage: "Nearly Half of All Men Say They Do Most of the Home Schooling. 3 Percent of Women Agree."

Further busting the healthy-heterosexual-marriage myth, at least for women, is the fact that its benefits are unequally distributed. The health boosts apply vastly more to men. Conversely, it appears that without a spouse encouraging healthy behaviors, many divorced straight men—at least those born in the middle decades of the last century—languish. They drink more; they smoke more; they exercise less. No one is there to make square meals appear magically before them or to arrange their social life. Granted, marriages that began 40 years ago were more likely to hew to traditional roles, but a dismaying number of modern marriages still cleave to them. This is one of the reasons that divorced men are more likely to remarry and to do it as bloody soon as they can. (Of course there are additional reasons having to do with other inequities, like the fact that men end up with more resources than women after divorce and women suffer the age penalty more than men on the dating market. The fact that men corral more resources is particularly galling in light of new evidence that wealth really is linked to happiness, meaning that a heterosexual woman's ex-spouse—assuming he remarries—is likely to end up happier than her, as well as more financially secure.)

There is some evidence that single women (especially those who have never married) actually live longer than married women. When I heard this I imagined them poorer but free from a life of drudgery, pursuing their crafts à la Jane Austen (oh, wait, she had plenty of tiresome chores and in fact died quite young). Women in heterosexual marriages tend to be the caretakers, and caretaking is brutal, emotionally costly, and often underappreciated work. But we know this. As *Dr Foote's Home Cyclopedia of Popular Medical, Social and Sexual Science* put it in 1900, "Marriage is no picnic for the housekeeper."

A study by the National Institute on Aging in 2000 examined personality traits in 2,200 men and women in their 40s at two different time points approximately eight years apart. It found that well-educated women who got divorced in midlife rated themselves as higher in gregariousness, activity, and fantasy after their splits. The authors speculated these slightly positive shifts in these particular traits were due to increased independence. The divorced men, on the other hand, showed modestly more depression, and less striving, self-discipline, and deliberation, especially if they hadn't remarried.

Learning about zippier, long-lived single women, and about the toll of being in meh marriages, I brightened. "So maybe it's good for unhappily married people to get divorced?" I asked, hopeful.

Uchino peered at me. "Not exactly," he said. "In terms of stressful life events, divorce ranks just under death." Then he peppered me with more discouraging stats from the divorce literature. Being divorced, it turns out, is particularly challenging for your health. In one 40-year study following 2,000 heterosexual residents of Charleston, South Carolina, those who were divorced at the beginning of the study (and stayed single) were 57 percent more likely to have dropped dead by the end of the study than their married counterparts and about 26 percent more likely to do so than the never-married singletons. Divorce was a greater risk factor for death than smoking. More recently, a large analysis of data from 6.5 million people in 11 countries found that people who are divorced are 23 percent more likely to die early than their married peers. Perhaps the surgeon general needs to stamp every marriage certificate with a warning label. *Use Caution Breaking Seal.*

Our friend William Farr, the physician from 1858, noted as much.

Divorce was rare then, but losing a spouse to death caused the untimely end of many a marriage. Regarding single people, he wrote, "If unmarried people suffer from disease in undue proportion, the have-been-married suffer still more."

Researchers now have data on the relative health effects of being single, being widowed, and being divorced. And the worst of all states, Uchino was sorry to tell me, is being divorced. It's even worse than being widowed, at least from the perspective of health. And compared to people who choose not to marry and who often happily navigate the shoals of singledom, divorce remains, in the words of one health study, "a costly life event."

The longitudinal health data is in: Being married can suck, although it's generally healthier than being single. One thing worse than being single is being formerly married. And the worst state of all is being formerly married when you wanted to stay married.

Why not just shoot me now? Uchino must have read my face because he did, finally, bring forth some better news.

For most people, he said, the ill effects of divorce dissipate over time. Health outcomes are worst during the first four years after separation, and they can disappear altogether if you remarry and stay (happily!) married for a long time. So if I was harboring any malcontented theories about the state of long-term love, here was another reason to subdue the cynicism. Falling in love again is recommended if you have any hopes of *staying* happily in love and thus thriving into old age.

I wasn't ready for any of this.

But I was also taking careful note of the worst scenario in case I never got ready. There is an exception to the divorce-recovery trajectory. Around 15 percent of divorced people simply don't get over the loss, not after four years, maybe not ever. As with some people who suffer decades of complicated grief over the death of a loved one, some people really can't pick themselves back up after heartbreak. These are the ones who skew the stats. They stay sicker and they die younger.

Learning about these outliers was like hearing about an emerging zoonotic disease.

Was it going to jump to humans? Was it going to jump to me?

MY CONTACTS AT the university suggested one more researcher for me to look up while I was in Utah: Paula Williams. Williams, a psychology professor, has spent much of her career studying the factors that make us more able to weather life's tough blows, like getting enough sleep.

Our conversation didn't start auspiciously. Adding to the chain of bummer trivia, she told me that while men's physical health takes a dive after divorce (if they stay single), splitting up is harder for women emotionally, especially if they didn't initiate it. "For women, breaches of major attachment and betrayal are a major stress, maybe some of the most stressful of all," she said, because women tend to be more relational and more focused on keeping close relationships intact. Men don't like being heartbroken either, but they may take blows to their job or financial status harder.

While Uchino studies the impacts of stressors on large populations, Williams is more interested in individuals. "Epidemiological research does not consider individual differences," she said, in a statement I found wildly hopeful. Data is not destiny. So why are some of us more resilient in the face of something like a breakup? Do personality traits matter? Early life trauma? The short answers are yes and yes. But she's convinced even these factors can sometimes be outmaneuvered. Ultimately, she wants to help us figure out how. If individuals can game the odds, then maybe I could, too.

Williams is a fan of the so-called big-five personality theory, the idea that the traits defining how individuals move through the world can be broken into five broad categories: introversion/extraversion, neuroticism, conscientiousness, agreeability, and openness to experience. It's generally assumed that these traits, like IQ, change very little over our lives, or at least not much until we are geriatric. Where we fall on the spectrum of traits can influence our health and behaviors. Conscientious people tend to get better grades and better jobs; neurotic people tend to be more anxious and depressed; extroverts can be more impulsive. All the traits have upsides and downsides. People high in neuroticism can be very motivated, for example, while people high in openness might be too easily distracted. But in general, openness is linked with many desirable qualities like curiosity, mental flexibility, and creativity.

Williams and others have also noticed that high openness appears strongly related to the ability to recover from stressful events. So what does it mean to be "open"? The trait is broadly characterized as comfort with novelty and desire for "cognitive exploration." To measure it, psychologists use the extensive five-trait questionnaire called the NEO (the abbreviation stands for the first three categories: neuroticism, extraversion, openness). The openness category breaks down into five clusters of questions designed to gauge imagination and fantasy, adventurousness, attentiveness to inner feelings, tolerance of others' viewpoints and ideas, and ability to appreciate and be moved by aesthetic experiences. People scoring high on openness really *feel* things, and they're tuned in to how they're feeling them. These are the tree huggers, the embarrassing burst-into-song people, the hopeless dreamers. It's basically like being a Pisces, or a William Blake (whose wife once said, "I don't see much of him. He's always in Paradise") or a Beethoven (who reportedly did in fact hug trees and once exclaimed, "Happy everyone in the woods. What glory in the woodland!") or a Walt Whitman, who, don't forget, wrote a paean to the glorious smell of armpits. You don't have to be a transcendentalist to be open; the label would also apply to many religious people, gardeners, cooks, explorers, meditators, scientists, lovers, travelers, readers, you name it.

Williams's lab was homing in on an important subcategory of openness: aesthetics—what she calls the ability to be blown away by beauty—and its relationship to managing difficult times. Her interest started 12 years ago, when she and her students were measuring emotional reactivity and blood pressure while having people recount stressful events. Everyone registered the stress response physiologically, but only people who ranked as highly aesthetically engaged reported positive, growth-oriented feelings about their experiences.

"If you're connected to art, nature, and beauty, you are periodically being forced out of yourself to think about connectedness to something bigger than you," said Williams, who exudes a calm, warm quality. I was sitting across the desk from her, admiring her classic-but-not-matronly sweater layering. "And if you can do that, then learning is better and understanding is better."

I leaned forward in my chair. Nature? Beauty? I was a fan before, but I was no longer just interested in merely feeling happier or more creative. Now I felt my very survival was at stake. I needed a whole new understanding of myself, and I needed to make sense of a world that felt upside down, scary, and lonely—and this was before a global pandemic rendered many of us feeling in need of similar balm.

Williams, who herself has been through a divorce, continued. "People who are open to experience are cognitively wired to grow and learn and move on, to do something that feels transformative."

Stacy Bare had said wilderness would help me. Now Williams was backing the idea with her data, and in ways that suddenly felt deeply relevant. But it also, I admitted, seemed a little far-fetched. Looking at beauty could change our self-image and help us transcend life's struggles? I'm not easily swayed by opportunities for transcendence, however carefully they may be wrapped in science.

I asked her to explain how this might work exactly.

She described one of her recent studies looking at brains-on-beauty. Williams and her team analyzed brain scans and personality data from 1,000 healthy participants in the Human Connectome Project, which aims to map neural circuits, structures, and functional connections in the brain and to see how individuals differ. The white-matter connections in the resting brains of high-open people were a little bit different in places, but where Williams saw consistently exceptional patterns was in people who answered yes to one particular aesthetics question on the NEO: *Sometimes when I am reading poetry or looking at a work of art, I feel a chill or wave of excitement.*

These people showed a marked increase in connections between parts of the frontal cortex associated with self-concept and parts of the brain associated with processing sensory and motor information. It's hard to know what to really make of this, and it's tricky to attribute emotions or insights to people based on functional brain images. But Williams's theory is backed up by some other research. Well-connected brains in these areas, she said, tend to be pretty good at processing stressful information and making narrative and personal sense of it. In other words, these drunk-on-beauty people know how to tell themselves a story when something confusing happens. The single emotion they share is awe.

A classic awe experience incorporates elements of fear or surprise to the point that when we are truly blown away by beauty or the power of nature, we may struggle momentarily to understand it. But, she said, "people who are very aesthetically prone are making a connection between the environment and their internal sensations and feelings." This sensitivity, she believes, translates into the ability to integrate stressful life experiences into what she calls "narrative coherence."

An earlier study found that viewing awe-inspiring videos of nature or childbirth (but not merely happy or comic videos) led to a "tendency to orient oneself toward a larger transcendent reality." Awe was the emotion that moved the participants to see themselves as part of a larger, meaningful reality, which is certainly a useful perspective if you're feeling singularly lonely.

I wanted some of that.

I was surprised, though, that not all brains respond the same way. We all like looking at beauty; we all tell ourselves stories. But apparently, some tales are more self-transcendent, or at least lean more to the thoughtful-yet-positive category after threatening events.

"Doesn't everyone get moved to tears when they hear incredible music?" I asked.

"No," she said. "Like when you describe aesthetic chill, they have no idea what you're talking about."

What she's talking about, to be clear, is a case of the goose bumps. William Blake appeared to be correct when he wrote in a letter, "The tree that moves some to tears of joy is to others a green thing that stands in the way." In fact, only about half of the 1,000 volunteers in the Connectome Project reported a tendency to get goose bumps when experiencing a peak emotional experience driven by beauty.

Was I prone to goose bumps? I was pretty sure I was. But I wasn't going to wait around for beauty to find me. Williams was positing something radical: that awe was tied to the trait of openness, and that, regardless of how prone to goose bumps we are naturally, we could learn to become more so. We could potentially shift this facet of our personality. If Williams was right, we could actively entreat self-transformation through cultivating the ability to see beauty.

Could I do that? I needed to do something. Uchino, with his research on the health consequences of heartbreak and divorce, had made clear just how high the stakes were. Now Williams offered a possible, life-giving corrective, one that was transformative and yet little known. For the sake of myself and so many other love-skunked people out there, I would see if it could work.

To claw my way through heartbreak, I would try to awe my way through it.

I knew one place to find it: outside.

5

OG SIN

You suck at love.
—SIMPLE PLAN, FROM THE ALBUM *GET YOUR HEART ON!*

I always liked driving west. After Salt Lake City, I'd flown back to Boulder and now I was headed out again to speak at another conference, this time in the tiny alpine town of Telluride. I picked up the old Ford Ranger truck we still shared with Peter and Lisa. Manhandling the stick shift and hand-rolling the windows up and down made me feel less ghostlike and more embodied: bigger and slightly tougher. Perhaps I was wandering alone through a dark wood, but I had my metal steed. This steed supplied no air conditioner, so the mountain wind slashed into the cab with the smell of something alive. Maybe this was the simplest tool of female power: a Ford.

The San Juan Mountains never looked better. As part of the conference one day, I led a group of hikers up a trail to a small creek. Some medical technicians joined us from town, and they brought blood-pressure monitors. We took the hikers' resting blood pressure before starting, and then again about 30 minutes into the hike. We had closed our eyes to hear the breezes and the birds better, crumbled and sniffed ponderosa pine needles, and felt the texture of the trail beneath our feet. It was fun to demonstrate in real time what I'd written about the quick restorative effects of engaging your senses outside. Sure enough, everyone's blood pressure dropped significantly. It reminded me again how badly I needed these lessons myself. I wanted to stop buzzing, to stop losing weight, to stop the ceaseless sorry soundtrack of heartbreak, and to remember I was not friendless and the world was not trying to kill me.

On the last day of the conference, I met a man named Ennis, a good friend of a friend. He was appealing in a nerdy way, confident, earnest. A systems scientist with a specialty in wildlife population dynamics, he wanted to save global ecosystems. He told me his wife had dumped him several years earlier. He was happy to offer advice. We were standing over a cheese plate at an outdoor party. A band was playing. Hipsters, film people, and extreme athletes were crowding the kegs.

Ennis told me the best thing about being newly single is that you get to fool around with other people. "The grass is greener!" he bellowed. He spoke in shifting tones of exclamation points and resonant warmth, an extrovert's extrovert. He was almost giddy recounting the freedoms and possibilities of singledom. I was mystified that anyone could feel so good about what was clearly a consolation prize. I liked being married. I liked the comforts and predictability, the rituals of couple-filled school meetings, family meals, dog walks, shared friends, the expectation of adequate retirement benefits. Happy coupledom was, as Bert Uchino kept saying in Utah, the holy grail of life as a naked ape. It was your best bulwark against chaos. Somebody had your back.

"You need to find someone at this party to make out with," Ennis said, his bushy eyebrows rising and falling above Scandinavian-chic eyewear.

I looked at him skeptically and scanned the crowd of sundresses and light fleece. We chatted a bit longer. Ennis migrated to a conversation with a polar explorer, and I ducked out to meet a friend for dinner.

As I walked to the restaurant, my phone rang. Ennis had gotten my number. "Did you find someone?" he asked.

"Everyone there was 24 and stoned," I said.

"Well, how about someone who's middle-aged and slightly buzzed?"

It took me a moment to realize he was offering himself. This had not occurred to me. I hadn't flirted with a stranger in decades. That muscle had withered, along with the inclination. Ennis, with those assertive eyebrows, those doleful eyes, was not an obvious magnet of ardor. But I was finding it fascinating that he was so brazen, and that he seemed to like me. I felt a gravitational pull toward his desire. This was wholly unfamiliar. And kind of sexy.

"Meet me in an hour," he said, and hung up.

But then I freaked out.

Um . . . I'm chickening out? Not ready, I texted him before our rendezvous, which was set for a pretty Telluride square bearing a notable cottonwood tree.

Um . . . you need to call the shots with more certainty to be convincing. Long hug under the tree in 36 minutes.

Why 36?

3x3x2x2

Now ur sounding like a systems geek.

That sounds like a yes.

I'm thinking.

What would 88% of your girlfriends tell you to do? The wise ones who really know you.

Compelling argument.

See you soon.

———

I LOOKED IN the restaurant's bathroom mirror, seeing a woman who'd once been sure of herself and relatively untroubled. I wasn't young, but I wasn't ancient. My hair looked smooth and well behaved in the Colorado aridity. Several weeks after one's marriage ends might seem like an impossibly short time to start snogging someone new, but something I hadn't felt in a long time was rising. It was part defiance, part survival instinct, part arousal. These were the feelings of youth, and I hadn't felt any of them in a long time. A certain teenage recklessness was knocking on my forehead. Plus Helen Fisher had said romance was good for us. Paula Williams said be open minded, open minded, open minded. Lisa wanted me to un-numb. *What the hell? Why not?*

A man was maybe seducing me, and I was liking it. And he was my age, and had a job, and was vetted by a trusted friend. I met Ennis under that cottonwood tree, lit by a near-full moon. We walked to a footbridge over the rushing creek. We kissed, a lot, and it was . . . amazing. It was almost a religious experience. No one other than my husband had kissed me like that in over two decades. And even he—not in a long time. I had thought I was very possibly done with men, and I was okay with that. I have good friends. I have nice kids, loyal pets. And yet, here I was, overwhelmingly turned on. My hard little heart hiccuped and started to soften, along with everything else.

I had discovered the power of sex as a teenager, but its power to me then was mostly a social one. It was a way to be in relationships, to know someone in a special way. That was so long ago I'd forgotten. I'd enjoyed it since then, fairly often, and then not so frequently, but for a long time it hadn't felt engulfing or piercingly intimate or even that important. It had felt like a wan houseplant. You have remote memories of blooms but they are distant.

After some kissing and gropes, I sent Ennis back up the gondola to his room, and I went back to mine, feeling humid and relieved it hadn't gone further. There was no plan to reconnect, and that was fine. I had a kissy souvenir from a fantasy town. Tomorrow Ennis would return to his New England city. I'd be back at my house-sit sending postcards to my kids, slightly flushed but still respecting myself.

Early the next morning, Ennis texted.

I should be with you in bed right now, my hand cupping your ass.

My first thought was, I'm going to have to put a passcode on this phone. My second was, How nice.

I was surprised that over the next several days, Ennis kept texting.

I want to see you.

He wanted to know my schedule, my upcoming coordinates. We talked on the phone. Sometimes for hours. He told me he was dating here and there, that it was good but not great. He was looking for a serious relationship. We acknowledged that we didn't seem like a practical pairing. We lived far apart. We both had kids. Still, he told me, he was interested. I was age appropriate, smart. He liked my body type. We laughed and pulled out the calendars. We had one clear spot in a few weeks. "I like Boulder," said Ennis. "I'll come visit you there."

Then I got nervous. I'd forgotten how to be sexy. I was still so raw from the split. Who wanted a girlfriend who cried every day? I had enough going on. I didn't really think I could handle this physically or emotionally. Plus, who was this guy, really? I knew he'd been on Tinder for years. I worried he was a walking petri dish of sexually transmitted diseases. I called my friend who had introduced us. "Is he really a good guy, from a girl perspective?" I asked.

"He's great," said my friend. "He's looking for something real. He has no red flags."

No red flags.

ON TOP OF all my other agitation, I now felt a surging river of carnality running through me. I listened to many podcasts, so many, often at three and four in the morning. I liked the ones about Eleanor of Aquitaine and the medieval drafting of the Charter of the Forests, because they sometimes put me to sleep. There were also lots about sex, how to do this and that and focus on your breathing and avoid STDs. It was all part of my crash course into unmarried sex. But it sometimes felt petrifying.

I walked around in a whiplash of fear, insecurities, arousal, distress. It all felt foreign. The arousal piece was astounding, like acquiring a lost sense, or installing cochlear implants in some unusual body parts. I thought about sex nearly all the time. I could detect subtle vibrations in roomfuls of people, in the supermarket, in radio newscasts. An underground current had become sensate to me. It was thrumming in me and around me. I wondered if certain people could see it like a curse or perhaps a blessing. I wasn't sure which.

Lisa hauled me off to a party. Some old friends were there. "Wow, you look *different*," said my (married) friend Greg, approvingly. He looked right past the sunken eyes and bulging ribs. He sensed something, like we had a secret handshake.

Ennis was flying out to visit soon. We'd been discussing logistics. He wanted me to line up a guitar he could borrow. He reported that he'd been given the "all clear" on a round of STD tests I'd asked him to get. "Good timing," I texted.

I was trying to stock the house with food he'd like. "Do you drink coffee?" I had no idea. "Chocolate: dark or light?" I knew nothing about him.

Boxers or briefs?

Boxer briefs. Right combo of loose and snug.

How amazing, I thought.

During a long phone call, I told him about how weird and hard it had been to say goodbye to the husband. He asked what I wanted in the new, post-marriage me, and while I was primarily thinking, sex, sex, sex, I didn't say it.

"A renewed interest and faith in men," I said. "Maybe you will help me with that."

"I already have," he said.

———

I DIDN'T HEAR anything from him the night before his flight, but in the morning he texted that he'd just barely made the plane. He was on the tarmac. A song popped up on the text chain, "Love Shack." I smiled and sent him a song back, "Like a Virgin," *Glee*-cast version. I bought more groceries. I changed into a blue-and-white dress, some freshly purchased underthings. I tied my hair up in a bun.

I stopped by Lisa's for a quick pep talk, and to borrow her Honda, which has air-conditioning.

"I'm a wreck," I said.

"But you look great," she said. "You've lost like 400 pounds and you're wearing a sundress and you look like Audrey Hepburn." I chose to gloss over the problematic association of extreme thinness with desirability and just take the compliment, because we both knew I needed it. Plus I'd spent some time on that bun.

"So why am I so nervous?"

"Well, there's the fact that you have been coupled for 32 years. And that you're about to embark on full-frontal vulnerable living at age 50."

With that, she hugged me and handed over her keys.

ENNIS ARRIVED CURBSIDE in a black T-shirt and black jeans. A large duffel. Sunglasses. He was going for the nerd rock-star look. Maybe he was nervous too. We kissed for a bit, both smiling, in the glaring, flat midday sun.

"Can I drive?" he asked.

"Okay, why?"

"I just like to drive."

Back at the love shack, he tried out the guitar. He strummed some Dylan and Petty. His voice was not great but he crooned earnestly, soulfully. I looked around and marveled. Here was a living, flesh-covered man, one who had gray eyes and well-developed quadriceps, singing to me in a borrowed house in Colorado. He was adorable. He had fallen from the sky, and he wanted to touch me. He put the guitar down. I straddled him on the couch. We made out and then moved to the bedroom. I was ready.

Apparently readier than he was. He cleared his throat. "All in good time," he said. I realized that meant that he didn't have an erection. Oh. "I guess I'm nervous," he said. We focused on me, and that was very nice.

He seemed tired. He'd had a long day.

"We can get some food?" I said.

"Yes!"

I got dressed in another outfit I'd thought about, long shorts and a patterned tank. We pulled some townie bikes out of the garage and rode to the farmers market. We picnicked on a blanket, our legs touching. Sometimes we held hands. This felt miraculous. I wondered if I would see people I knew, people who didn't know I no longer had a husband. I wondered what they would think. Just then, a kid I knew, a friend of my son's, wheeled past us on a skateboard and waved. Okay, that was weird. It was all weird. Every damn thing was weird.

WE HIKED THE next morning, came home, showered, and fooled around some more, but his body was still not cooperating. Maybe it was the altitude. More likely, I worried, it was something about me. We grilled some steaks and sat on the porch. We monkeyed with a portable speaker, and finally got it to connect to his phone. "You are now paired," it announced in a flat electronic voice. I smiled and poured the wine. I wasn't feeling very paired. But I was hopeful.

Later that night, he told me what to wear: a black teddy or tank. "Hmmm," I said, my mind going over what I had. "I have a black T-shirt?" He nodded. I didn't really mind him telling me what to wear. I found it sexy that someone even cared what I wore. I wasn't surprised when he fitted me with a blindfold. He'd hinted about his proclivities, and I was curious enough to give it a go. He led me into the bedroom. He asked me to put my hands on the wall and bend over. He said it in a mock-gruff way. I giggled. "It's not funny," he said. I bit my lip. Reader, some spanking ensued. He turned me around and attached something to my nipples, under the shirt, so that they stuck out against the fabric. Clamps? I hadn't known they were a thing. I pictured the rigid plastic ones I used to reseal the kids' pita chips.

"Here's the safe word," he said, and he whispered it to my ear below the blindfold elastic. It was his last name. I couldn't see anything, but I could imagine how things were coming out of his duffel bag, Mary Poppins style. When he put me on the bed, there were ropes already on it. He attached three of my limbs to them, expertly. I could hear him undressing. Then he straddled me, mashing my breasts, squeezing them around the pita-chip clamps. My bottom stung. This is a lot of effort to get an erection, I thought. I heard the buzzing of a vibrator, and he pressed it against my pubic bone. I tried to shimmy into a better position, but he was handling it like a pestle. Meanwhile, the clamps were feeling like the jaws of death. At some point, I coughed out the safe word, in a nice sort of way, with an um and question mark at the end. Ennis adjusted the paraphernalia around my face.

"Hey, can you just use that thing a little more lightly and a bit lower?" I nodded and smiled, so he'd know we could start up again.

He sighed. He rolled off me.

THE NEXT DAY, after hiking, I showered and dried off in the bathroom. Ennis came in and hugged me from behind. He kissed my shoulder. We looked in the mirror. "Did I do that?" he said, seeing the bruises on my breasts. "Sorry. How's your bottom?" He looked. "Ooh." Also bruised.

"Why did you tie up only three of my limbs?"

"So I could flip you over."

Ennis was clever, but not, perhaps, adept. I was willing to test the idea of surrender, to be told what to do at a time in my life when I had no idea what to do. This was supposed to be a chance to learn a few things. But what had happened the night before wasn't what all the podcasts had prepared me for: good sex, they all said, was supposed to be about communication. He didn't really expect me to just be quiet, did he? I was willing to try again. We could learn together! His intentions were good, I thought. He'd come out here for *six days* to try us out. I wanted to believe he wanted a real connection too.

We worked some on our laptops, talked a little, made snacks, held each other at random times during the day. I told him I was taking gummy bears to help me sleep.

"Oh my God, this is Colorado!" he said. "Forget sleep, let's go get something fun and euphoric."

So we did. OG Sin: it smelled skunky and exciting.

That night, we went out to hear some music with friends. They found him funny and smart. He was affectionate, his hands roaming gently over me. His body swayed against mine. I found the full physical contact delirious. My husband had never been touchy this way, and so I never was either, but I was liking it.

Over the ensuing days, we held hands all over town and hugged in the kitchen and threw our legs across each other on the couch. We were great talkers. About everything, except about the fact that we were no longer making attempts in the bedroom.

He would look at me fully, smiling, pouting, playful, a man-boy in soft T-shirts that smelled like grapes. He had advice, insights, a swerving way of laughing uproariously and then immediately being very serious, all eyes on me, all the time. He offered an unfamiliar zone of warmth, attention, and disclosure.

"What are your biggest fears?" I asked him one afternoon on the couch.

"My biggest, biggest fears?"

"Your big ones or your small ones. How about a combination plate?"

"I think my biggest fear is not mattering. I want to have an impact in the world. I want to be part of the people in my life. Part of their lives. I want to create the world more towards my vision of it. I want to matter."

He was such a hard-charging white guy, like my husband, in ways I found both compelling and dispiriting. Women generally don't say things like that. We are used to not feeling like we matter to the world stage or that we are entitled to bend it so easily to our liking. And it's a given that we are part of our loved ones' lives. That is our main currency.

"What are your fears?" he asked me.

There was so much I could have said. Of growing old and invisible. Of losing myself and not finding her again. Of dying single and broke, like my mother did. But what I said was also real.

I told him I'd been thinking a lot about intimacy, because that was one of the things that I was confused about. I wondered if I had trouble accessing it. This is one of those legacies of a broken marriage, as you account for all the things that you did wrong or that you're not good at. "I sometimes wonder, is this my issue? Do I have a hard time with intimacy? I don't know," I said. I was afraid that I didn't know how to give or receive love, that my heart had grown anesthetized.

It was strange to be talking to him about this, but we were lurching toward some kind of intimacy in our way, so it was also not strange at all. He was here. I was here.

We were both cloaking and uncloaking our vulnerabilities all week.

I WAS CONFUSED. During the day, we were nuzzly and affectionate. But Ennis was supposed to be my priapic Adonis. He was supposed to ravish me and relish me and help me regain my sexual confidence. He'd signed up, with brio, for that exact job description. I was disappointed we'd failed to connect in an important way, and I was rattled that it must somehow be my fault. I now felt worse about myself, my attractiveness, and my future romantic prospects.

I'd read about men's testosterone levels being highest when their partners are ovulating. I was two weeks out from that, so maybe I was emitting some kind of pheromonal Weedwacker. Or maybe women over 50 really are unfuckable. Or maybe just I was.

On night six, our last night, we were drinking wine and smoking a little on the porch.

"Can we talk about what happened here?" I asked. "I feel like I need to understand it."

"Okay," he said. He didn't look happy but he looked resigned. He shrugged. "I'm just not feeling the chemistry."

"But we had it in Telluride."

He then explained some stuff that was hard to hear. He said it was a turnoff that I'd asked him to get an STD panel, and another turnoff when I laughed and used the safe word during our brief dungeon experiment.

"You were," he said, "a boner killer."

"Boner killer? I was a boner killer?" I knew this phrase might never leave my brain.

He winced. "Look," he said, "I usually date women I meet online, and I have in my profile that I like to be dominant, and so I match with women who put down they like to be submissive. You are so different from the other women I date, and I really wanted to try." He was quiet for a moment. "I think I need to learn how to be with a woman I really want to be with."

"So, how many women *are* you dating?" I was trying very hard to keep the tone light because I really wanted to know.

"Right," he said. "Okay. Well. So there are, like, between four and six other women, depending on how you define it."

"Between four and six."

"Yeah."

It took a moment for me to process this. I flashed to a scene in Spike Jonze's film *Her*, when the main character, a man who has fallen for a honey-voiced operating system named Samantha, learns that she is simultaneously conversing with 8,316 others and has declared love to 641 of them.

I was trying to breathe. "How do you keep them straight? Isn't that confusing?"

He laughed. "Yeah, you don't want to mess it up. I take notes in the back of a notebook. Names of their kids, stuff like that."

"Do they know about each other?"

"Well . . . it's like what I told you. I say, *yeah, I'm dating.*"

True-ish. Ennis did say that, but he'd also told me he was looking for a serious relationship. Didn't he?

"What if you really like one of them? Then do you see more of her?"

He seemed to be appreciating my interest. This wasn't something he got to talk about.

"Yes! Sometimes one can be sort of like a queen bee. The rest are the harem."

"Harem?"

"Yeah. I love it that I can talk to you." He squeezed my knee. He was thoroughly enjoying himself, despite a problem with mixed metaphors.

"But they don't know they're in a harem?"

"Not exactly."

WE BROUGHT IN the dishes. The walls were undulating. I wasn't sure if it was the weed or the conversation. I needed to be outside. We went for a late-night stroll.

"Do you have a queen bee right now? Maybe you're feeling guilty about being here with me and that's why you can't—"

"Nope. No queen bee. I'd like a queen bee."

He reached for my hand, as usual, and held it. "You're turning me into a pothead," he said.

"So, wait," I said. "When you take my hand, I feel a jolt of electricity. But you feel . . . nothing? Like we're having basically a fifth-grade romance where a boy and a girl hold hands and there's no hormones and that's it?"

"No," he said. "It's not like that. I do feel something. It's hard to explain. Everything is dynamic, and inputs change, and who knows what's around the bend for us or anything else? I *really* like you. Maybe middle-aged sex is just complicated."

That much was becoming obvious.

Still, I couldn't get over the fact that this was a man with a systems fetish who screws five women at once. He was a freaking alpha chimp, or a murder hornet, or something in between, and yet he couldn't get a hard-on for me. My brain was shouting one word at me: REJECT. Marital reject. Sexual reject. Boner killer.

We walked back to the porch.

The Siri-in-the-speaker sensed my phone and pierced the night.

You are now paired.

PART TWO

ALONE

6

ALL PAIN IS ONE MALADY

Rejection

She's making friends, I'm turning stranger.
—DAVID BERMAN, FROM THE ALBUM *PURPLE MOUNTAINS*

William James, the nineteenth-century psychologist who was said to write like a novelist (his brother was Henry James, the novelist who could write like a psychologist), floridly described social rejection, likening it to feeling essentially unseen and immaterial. It sounded, alas, like a portrait of a dying marriage. "If no one turned round when we entered, answered when we spoke, or minded what we did . . . and acted as if we were nonexisting things, a kind of rage and impotent despair would ere long well up in us, from which the cruelest bodily tortures would be a relief; for these would make us feel that, however bad might be our plight, we had not sunk to such a depth as to be unworthy of attention at all."

It's true that being flagellated by Ennis, both metaphorically and, er, literally, had made me feel more alive. But his visit was, by nearly every measure, an epic failure. I had put myself out there, disrobed, revealed secrets. In return, he told me that he had a harem and, moreover, that I didn't make the cut. Also this: the bad cosmic joke that my rebound superhero couldn't get it up.

Lisa was waiting for me over turkey sandwiches. She'd been apprised of all the details, and she was not amused. She lambasted me about how nice I'd been, how I should have dismissed him at the first flash of nipple clamp, or surprise-harem clamp, or premonition-that-he'd-slept-with-another-woman-the-same-morning-he-flew-to-Colorado clamp.

"This wasn't some humorous lark," she said. "It's going to cut you."

I nodded. My eyes burned. "It hurts. I feel like I just took 10 steps backward."

"Right. If he wants to be, I don't know, sexting or domineering someone on the internet, I don't blame him. That's who he is. And so this is you learning to navigate this world."

But I didn't want to navigate this world. I didn't ask for being single. I didn't want to know this was what it was going to entail. And now she was bringing up the internet, the darkest of all possible seas. "I feel like everyone's out there comparing you to everyone else," I said, "because they're all sleeping with six people at once. I'm 20 years older than everyone else I'm being compared to."

"Is he dating people 20 years younger than you?"

The harem, he told me, skewed young. "Yeah."

She looked mad. But she was mad at me. "Look, he's got some stuff about sex and women. But it's up to you to be the guardian of you." She rattled her sandwich at me. She went on. "Your marriage was a long, slow starvation diet and now everything seems delicious and it's not. Maybe you need to send that shit sandwich right back and say, Look, with respect, I ordered the wrong thing. I actually wanted a nice meal."

———

LISA WAS HELPFUL, as always, but I called in the experts. I had two phone sessions, first with my physician, who'd run a bunch of labs before I'd left town, and then with my therapist, Julia. My physician was alarmed. My blood sugars were weirdly high. One of my inflammation markers was also high. My gut bacteria were completely out of whack. I wasn't making enough pancreatic enzymes and therefore I was not fully digesting food. "Your body looks like it's in fight or flight," she said. My body was acting like it had been left alone on the savanna and was being circled by hyenas. She told me to take deep breaths before eating, to get more sleep, to take daily enzymes, and to visit a specialist when I got back to town.

My therapist always seemed pleased to see me, even on Skype, as if I were her most interesting patient of the day. Fleeing husband? Sado-masochistic lovers? She was all in. I'd been seeing Julia since my husband announced he wanted to pursue a soul mate about six months before he moved out.

Julia pointed out that I was feeling very fragile, and that's why I was taking the week's failures with Ennis so hard.

"There's a lot about him that is compelling and intriguing and fucked up in ways that might take years to sort out," she said.

"Why does it all hurt so much?"

Much of it, she said, was timing. "You don't feel self-confident. Until then, dating is going to be a bit bruising."

THERE'S A DIRECT line between heartbreak and fragility. The words are etymological cousins: broken shards all around. When we feel unloved by key people in our lives, we easily assume we are unlovable. As a field of study, rejection and ostracism came to the attention of psychologists a bit late in the day, which is surprising considering that they can so dramatically alter one's behavior. There are now whole subcategories within psychology specializing in social rejection, sometimes called social exclusion, or ostracization, as well as partner separation, partner loss, and, of course, grief. Heartbroken, we are likely to hit all the emotional bases in more or less this order: rejection, grief, shame, and existential loneliness, with each one changing the brain in ways that create anguish and often distinctly self-defeating behaviors.

Kipling Williams, a social psychologist at Purdue University, attributes the field's recent popularity to an unfortunate trend: mass shootings, in which perpetrators are often found to be social outcasts, spurned or just ignored by peers and romantic interests.

To study how people act when they feel rejected, Williams's lab designed a classic psychology experiment known as Cyberball. A subject is told she will be tossing a ball in a computer game with two or three other players. The game starts out as expected, but soon the other players toss the ball only to each other, leaving out the subject. In a 2007 review paper of Cyberball and similar experiments, Williams wrote that feeling rejected in this way increases blood pressure and raises cortisol levels while "reducing feelings of belonging, self-esteem, control, and meaningful existence."

Cyberball isn't nearly as cruel as another experiment known as the life-alone prognosis paradigm, developed in the early 2000s by social psychologists Roy Baumeister at Florida State University and Jean Twenge at San Diego State University. In this one, researchers give subjects (typically college students) a series of personality questionnaires, supposedly analyze them, and then deliver a prophecy (randomly assigned): either the subjects are told they will likely enjoy rewarding relationships throughout life, including solid friendships and a long, happy marriage, or they are told they are the type destined for unfulfilling friendships and multiple or poor marriages, and they will likely end up alone. As a negative control, a third group is told they are accident prone and may suffer many physical injuries. Armed with the pronouncement of the oracle, the subjects then run a gauntlet of cognitive and behavioral tasks. The results are dramatic. The future-loners group, for instance, performed the worst on the general mental-ability portion of the Graduate Record Exam and on a logical-reasoning test, and they exhibited the worst self-regulation by eating more unhealthy snacks. Their higher-order cognitive thinking had taken a big hit. In a similar experiment, brain scans revealed that the social losers showed less activation in the areas associated with attention and executive functioning, leading to poor judgment.

Wrote Baumeister, "Some loss of self-control is one of the negative effects of social exclusion." In other words, rejected lovers often do things they may later regret.

Being given a bad-news prophecy wasn't universally damaging. It was only the social bad news—not the expectation of accidents—that changed behaviors. (Don't worry: the would-be loners are not destined to binge on cookies for life. At the end, the experimenters revealed the bluff.)

Formal ostracism has been around since at least 500 BC, when the ancient Greeks practiced *ostrakismos*, voting on whether or not to banish certain fallen political leaders for 10 years. Peer-sanctioned ostracism has been around much longer and is found across cultures and even across species. Chimps may banish other chimps who repeatedly put their own self-interest before the group's; elephants have been known to isolate individuals for breaching codes of behavior. Scientists have looked to evolution to explain it. As Kipling Williams argues, the act of expelling wayward individuals serves to strengthen the cohesion and thus survival of the remaining group.

For the outcasts, such punishment can be fatal, or it can be helpfully corrective. Human parents give children "time-outs" because they are powerful, effective ways to enforce behavior. Most kids naturally want to rejoin the family zone. Lovers and friends may give each other the silent treatment; men and women ghost their recent hookups with surprising regularity. *Dismissed*. Social media creates all kinds of new opportunities for exclusion and misery. With one click you can become blocked, unfriended, unsubscribed, unfollowed, deleted, canceled.

Spurned lovers and friends don't always take it well. Rage, vengeance, jealousy, and desperation are not qualities that mix well with impaired self-regulation. Hell hath no fury . . . Cleopatra uncorked the asps, Richard III executed everyone who gave him the hairy eyeball, and beastly Heathcliff just kept locking women up. We applaud when Odysseus slaughters his wife's suitors. They deserved it. But when Othello takes out Desdemona, not so much.

One attempt to study vengeance brought on by rejection is known as the Hot Sauce Test (who comes up with these?). In a rather elaborate experiment at Northeastern University, participants (presumably heterosexual) were told they would pair up with a partner for some problem-solving tasks. They were also told there would be a separate taste test (in this part, there was a subtle opportunity for them to learn that their opposite-sex partner hated spicy flavors). Halfway through the problem-solving part, the partners (who were actors) told the participant either that they would prefer to work with someone else or that they had to run off to an appointment they forgot about. The research subjects were then asked to prepare the actors' taste tests for later and were given a container they could fill as high as they liked with hot sauce that the actors would be required to swallow in its entirety. In the scenario in which the male partners left a female subject to work with a rival, the spurned women reported lower self-esteem and greater feelings of jealousy than the participants whose partner had left for the appointment, and they poured 50 percent more "fiery-strength" hot sauce into his container. When the genders were reversed, male subjects poured *four times more hot sauce* into their partners' containers. As the researchers noted, the gender discrepancy might just reflect a portion-control issue for dudes, but it's also consistent with the fact that men are far more violent in relationship conflict, with over half of all American female homicides committed by vengeful men.

If evolution provides insights into why we reject, exclude, and ostracize each other, it may also explain why it hurts so damn much to be on the receiving end. Since early humans' lives depended on not being kicked out of a group, it would make sense that we learned to be very, very sensitive to any signs of our own misbehavior or to others' mounting judgment of it. "An ostracism-detection system . . . probably coevolved with the widespread use of ostracism," writes Williams.

What would a powerful detection system look like? A hypersensitivity to social cues (especially negative ones) from other people, a keen sense that perhaps we're flawed or have done something wrong, and an acute capacity to feel shame. And perhaps the strongest of all: pain.

Our ancestors who felt those things most gravely were the ones who quick-stepped their way through the minefields of living in groups. They successfully reproduced, raised children, and passed their genes on to us, the easily heartbroken.

WHILE I WAS staying in Boulder, I stopped by the office of Tor Wager, a psychologist who runs the Cognitive and Affective Neuroscience Lab at the University of Colorado. He's developed a specialty studying placebos and the way the brain processes pain. Raised a Christian Scientist, he's long been fascinated by the way people's beliefs about healing influence the sensations and outcomes of disease or injury. "I grew up with people with a strong set of beliefs," he said. "I have a long-standing interest in what we can do with our thoughts and beliefs that can make a difference." Once, as a boy, he developed a rash and was given a random nonmedical cream. The rash went away, as did his mother's anxiety over the rash, probably just because they both believed the cream—and God—would make it better, he said.

Wager had no particular interest in social rejection. He just wanted to conduct experiments that made people experience pain so he could test placebos. He's scanned people's brains while their body parts were subjected to heat, shocks, and thumb-squeezing pressure. Among other things, he's been able to locate where in the brain people process pain and how much relief they experience even when given fake medicine that helps them release their own natural opioids. He's also interested in the emotions attached to feeling pain.

"We started doing rejection studies," he said, "because that's where there is a lot of emotion around pain. In a scanner, it's important to feel something."

And it wasn't just any social rejection they tested. Wager didn't want to look at Cyberball; he wanted to image Big Pain. He wanted to look at heartbreak.

Like most of us, he'd experienced it himself. Now in his mid-40s, he looks younger than his years. He's tall and loose limbed, has a scruffy goatee, and rides a bike to work. He's happily married now, but he remembers well the first time a serious girlfriend broke up with him in graduate school. "The grieving process is a natural thing, but it feels like it will last forever," he said. "I cried a lot. I didn't date anyone for close to a year."

To recruit candidates for the study, his team posted ads on Facebook and Craigslist, asking for volunteers who'd been dumped within the past six months. The volunteers slid into a functional MRI scanner while looking at a photo of their ex. Researchers compared the brain scans to those of subjects viewing a photo of a neutral friend and to those experiencing either a little bit of warmth or "noxious thermal stimulation," the equivalent of spilling hot coffee on their left forearm.

What the team found, like Helen Fisher's group a year earlier, were fireworks in the dorsal anterior cingulate and the insula during both the heartbreaker-photo-viewing and the scalding-coffee agony. The images overlapped. Wager's team wasn't the first to view the neural overlay between physical and social pain centers, but it was one of the first to try to quantify it. "In short," the team concluded, "intense social rejection activated somatosensory regions that are strongly associated with physical pain, which are virtually never associated with emotion as typically studied." Or, as Antiphanes put it more simply in 400 BC, "All pain is one malady with many names."

Next, Wager asked what seemed like a glib question. If social pain shows a similar neurological fingerprint to physical pain, can it also respond to a placebo? Other studies had shown that people who take common painkillers feel markedly better when they're rejected in Cyberball. Could it really be that easy? I considered running straight home and popping a Tylenol, but then I read some studies indicating these painkillers might also numb us to empathy, so this deserved more consideration.

Maybe I could get away with a placebo. Because Wager ran another study with the photos of the heartbreakers and the heat probe and the whole bit. This time he gave all the subjects a hit of a do-nothing salt-water nasal spray. He told half of them it was a special medicine to alleviate pain and sadness, and he told the other half it would simply help the researchers better interpret the MRI images. Lo and behold, the placebo subjects who were told it would lessen their pain showed less intense heartache in both the brain scans and in their own descriptions of discomfort from the heat and from looking at their exes.

Since Wager's and Fisher's earlier studies, new ones by him and others suggest the overlap between physical and social pain is not perfect. The individual circuits appear to be wired differently, or at least in ways that are too vague to be neatly characterized. What the newer imaging does confirm is that our neural networks take social pain very seriously. It shows up in consistent spots (in the anterior cingulate and elsewhere) that are associated with deep human survival in circuits of our brains that have been carefully preserved for a long, long time.

Not only does heartbreak "hurt," the pain appears to be there for a reason.

7

HEARTBREAK HOTEL

Grief

Anything dead coming back to life hurts.
—TONI MORRISON, *BELOVED*

Ennis had been gone a few days, but his texts kept appearing.
I miss you, he wrote.
Really?
It's complicated.

THE DRIVE TO seek romance, even of temporary duration, is one of the strongest propulsions we have. Good people will argue about whether we are innately programmed to desire nonmonogamous relationships, open relationships, or, in Ennis's case, a harem. I find myself drawn to Helen Fisher's description of our urges. Based on her wide gathering of cross-cultural data on the topic, she believes most of us do seek a secure, primary partner, but perhaps not forever, and sometimes we stray. And plenty of people can find happiness in other configurations of love, partnership, and solo-ship. The human brain, even when it comes to love, is highly flexible.

Nevertheless, our capacity for pair-bonding is one of the central defining characteristics of being human. In the animal kingdom, monogamy is quite rare. It is believed to exist at a rate of between 2 and 10 percent (it's up to 90 percent in birds, but straying is not uncommon). For the animals whose brain systems evolved in tandem with the practice, the rewards are high. In monogamous titi monkey males, for example, finding a long-term partner increases the brain's metabolism of fuel in the form of glucose, especially in the regions responsible for motivation, social behavior, and social memory. Beavers occasionally philander, but generally stay paired until death, the better for building and maintaining lodges and dams. Shingleback skinks of Australia can stay together for 20 years, and so can termites, whom we may not think of as paragons of romance. Staying with the same partner increases their defenses against predators. Alas, it's not all bliss; male and female termites will sometimes flee their comfortably rotting log if a better mate comes along.

AS THE DAYS turned into weeks, sometimes Ennis would call and we'd talk for a long time. He wanted to know how I was doing; he wanted to know if I thought he was a bad person, or just bad to women, or was he maybe still good for women? He wanted to think of himself as a Boy Scout, leaving the campground better than he found it. Um, no, I said, you are no Boy Scout. These calls I both anticipated and dreaded. I told Julia, my therapist, about a conversation in which Ennis admitted he'd slept with a woman the same morning he flew to Colorado to see me.

"Who does that?" I asked.

"This is not a moment for silver linings," she offered, "but you did learn that you have feelings, and they'll be available for someone else."

We discussed the wisdom of my taking a break from dating. This is, after all, the standard breakup advice. She said the last thing I needed was to be disappointed or hurt again. She wasn't doctrinaire about it, just cautious.

"It's hard to jump into life and strange relationships without risk," she said. "You're going to bounce around before you find someone. It's risky."

We agreed that my metabolism didn't seem cut out for any more risk at this moment.

"If you have some time as a single person, your nervous system will calm down and you will come to feel you're okay without anybody," she continued. "You have a recovery to go through, and you can't really recover until you feel confident and independent. It takes months. It's hard to do if you're infatuated with someone. It just postpones it."

It all made sense. But I found myself bargaining hard. How about a casual fling? No infatuation allowed. I ran through many variations and configurations of fling in my head. But I had to concede that Julia was probably right. I was in no condition to pursue love, and maybe I shouldn't have tried at all.

I remembered when, damp from sprinkler spray, Helen Fisher had exclaimed, *Have fun! Just don't get dumped again.*

Heartbreak is a beast wagging a long tail. It can make you more insecure, more likely to make poor decisions, and more prone to behaviors that are bad for you. Once the grooves of rejection get etched into your psyche by a disappearing partner, it is all too easy to gouge them deeper. Turpitude, stupidity, poor judgment, it's all right there, documented and quantified in the brains and behaviors of people who have experienced abandonment, loss, and the natural consequence of those conditions: loneliness.

Ennis rattled me to a degree clearly out of proportion to the time we spent together, this second romantic rejection coming so soon on the heels of the first. Despite all his fecklessness, I considered going to visit him and trying again. A disturbing question formed: What felt worse, that he had a twisted-sex harem, or that I wasn't admitted? Just the fact

that I was even asking it shows just how far I'd fallen into madness or desperation or self-erasure or some combination of all three.

I was in trouble. I was losing myself, and it scared me. I needed to
know what the heck had happened to the woman I used to be, and how
to get her back. I knew I would have to keep trying to understand more
fully what was happening to my brain.

THE UNIVERSITY OF Colorado, it turns out, is a kind of Heartbreak
Hotel, and not just because of wild football weekends in which frat
houses routinely burn couches on the Hill, the neighborhood abutting
the campus.

One building over from Wager's on the pretty, red-roofed Boulder campus sits Gold Biosciences. Bland, labyrinthine hallways course
through the department of molecular, cellular, and developmental biology, eventually yielding up the lab of Zoe Donaldson. A behavioral
neuroscientist, she is interested (like Wager) in seeing the signature of
heartbreak in our brains. But her ambition is to map it on an absurdly
granular level, at the scale of cellular neurons. You can't really go around
sticking mini-cameras in the heads of heartbroken people, so her subjects
are prairie voles. Perhaps you have heard that prairie voles are helpfully
elucidating the neurochemistry of love, attachment, and monogamy.
(Donaldson herself did postgrad work in the famous labs of Larry Young
at Emory University, where the magic love recipe of oxytocin, vasopressin, and dopamine is being carefully distilled.) But you probably haven't
heard about vole heartbreak.

If Wager's metaquestions are about pain and placebos, Donaldson's
are about grief. What is happening on a cellular level when we experience
heartbreak, and why do some of us seem to take much longer to get over
it? These are the 10 to 15 percent of divorced people who just get sicker and
more depressed as time goes on, and the roughly same percentage who
experience what psychologists call "complicated grief" after the death of
a loved one.

When Donaldson heard about these people, for whom grief becomes a long-term disabling force, she thought it might provide a good opportunity to seek insights from lab animals whose genes, social brain structures, and hormone receptors are remarkably similar to ours.

"We're interested in what's so special about grief and how grief is different from depression," said Donaldson, who greeted me in her summer nonteaching attire: leggings, T-shirt, coffee thermos in hand. Young and blonde, with a wide, open face, she looks like she could be gearing up for a spin class rather than managing a $1.5 million innovation grant from the National Institutes of Health. Summer is her favorite season because with students scarce, she can devote herself to the research, which involves not only running experiments on voles but also breeding them. Unlike with mice and rats, you can't just buy voles on the internet.

"What I want to understand is not necessarily what happens when you lose someone, but how do you adapt to that loss? Because this is the process that sort of goes awry," she said. I wanted to know these things too.

Complicated grief, sometimes called persistent bereavement disorder, is characterized by prolonged sadness, yearning, and longing, a sense of disbelief and difficulty accepting the death of a loved one. People with this condition may compulsively avoid reminders of their loved one and yet also be stuck ruminating over aspects of the loss. It's as if the loss just happened, over and over again. People tend to get diagnosed if they still feel acute grief after 6 to 12 months, but it's a controversial label because some losses, particularly the death of a child, are understandably debilitating for many years. Still, most grief is expected to lessen over time, to enable those left behind to regain some periods of joy, fulfillment, and normal functioning. At what point does normal grief slip into pathology?

Donaldson thinks complicated grief isn't exclusive to bereavement. "I have a feeling it also happens in divorce, and there it's emotionally a lot more difficult in some ways because you're also dealing with feelings of betrayal and knowing that the person is still there but you can't have them." I was curious what she could see in those little vole brains. It's not for nothing that vole is an anagram of love.

Prairie voles may not get divorced, but they know partner loss. That they even have partners makes them highly prized among scientists. Like us, prairie voles tend to be socially monogamous (meaning they shack up with a partner and raise their young together, with father involvement), but also like us, they are not strictly sexually monogamous. In other words, they sometimes wander. And like us, some adults prefer not to shack up at all and would rather just play the field.

In the wild, life for a prairie vole is short and intense. Resembling small hamsters, they are plump and apparently delicious. "They are the popcorn of the prairie," said Donaldson. "Everything eats them. Snakes, coyotes." It's not uncommon for Mom or Dad to pop out of the nest for some foraging and never return home. Their lives last perhaps several months.

Prairie voles are even a bit more loyal to the idea of coupledom than we are, at least these days, in late-stage capitalism. Once paired up, 75 percent of them will stay together until one dies, even when the female isn't reproducing. If the male dies, the female will rarely lasso up a new mate. If she dies first, 20 percent of males will pair up with someone new. Prairie voles are inveterate snugglers with their mates and their pups. They even console each other during stressful times, piling on top of each other, hugging, nuzzling, licking, and grooming. On the internet there are pictures of voles hugging and they slay me. Within a few days of their first mating encounter in a lab, males and females will overwhelmingly prefer to spend time with their lover over all others, even when sexy newcomers are dangled like taffy before them.

In Donaldson's heartbreak lab, the voles live in neatly stacked boxes made of polycarbonate. Fluffy and dark, they dart in and out of PVC pipes, do little chin-ups, and scrabble about amid piles of shaved wood. Donaldson assures me that her voles are quite happy overall. They reproduce well, get snacks, and live much longer than in the wild. She also tells me they are not bred in vain, that each one will contribute to science. She's currently breeding about 300. Half are monogamous prairie voles, and half are their genetic cousins, meadow voles, who happen to be not monogamous at all. They don't prefer one mate over another; they don't even much like hanging out together. This burl in the family tree is a great boon to scientists like Donaldson, because by comparing the two cousins' ever-so-slightly different brain structures and neurochemistries, they can learn about the unique molecules of paired affection.

Here in the Heartbreak Hotel, all marriages are arranged. She pops an unrelated adult male and female into the same cage, they sniff around a bit, and the male struts around, causing the female to start ovulating. Touching releases oxytocin in both partners. One thing leads to another. When they mate, stimulating her genitals and cervix, oxytocin floods from one part of her brain onto receptors in another part, virtually the same way it happens when she's giving birth, priming her for dewy attachment. His brain has more receptors for vasopressin, a neuropeptide that triggers behaviors of territoriality, mate-guarding, and devotion to home life. Pretty soon, they are inseparable. In their cages, they make full body contact much of the day. They huddle, in official lab parlance.

Then, as in a Greek tragedy, the Fates intervene. Donaldson parts them. Poor *Microtus ochrogaster*. From here, they tumble into one of several life narratives, otherwise known as experiments. In some, the bereaved partner can mate with a new vole. During "partner preference tests," a vole might be placed in the middle of a large cage. In one wing lives the long-lost partner, now returned; in another, a new potential mate. Donaldson and others have observed these reunions at 48 hours, two weeks, and four weeks post-breakup. Will the vole, as she puts it, "prefer Wife Number One or Wife Number Two"? (In case you're interested, and I certainly was, by four weeks, the voles overwhelmingly prefer the new mate. This pained me, I'll admit, but also gave some cause for hope—at least in voles, second marriages can work out.) In another experiment, the voles learn how to press a lever that will lift a door offering a reward, like some tasty Purina rabbit kibble. Then, one day not long after separation, the lost lover is behind the door. *Eureka*! The bereft vole will eagerly press the lever to reunite. But then Donaldson makes the experiment harder—now the vole has to press it two or three or four times to lift the door. Donaldson might once again remove the lost lover, so that one day, the *objet d'amour* is no longer there at all. How long will the lonely vole keep pressing the damn lever?

Remarkably, what Donaldson seems to be zeroing in on is an essential element of grief: yearning. "We think this is a proxy for incorporating the finality of the loss," she said. How hard is the vole willing to work to lift the door to be with his mate? And how long does it take for "acceptance" to set in that she is no longer there? Moreover, what is happening in their brains as these decisions play out? Through a fluorescent sensor implanted in the voles' nucleus accumbens, a part of the brain associated with emotional learning and addiction, Donaldson can actually watch individual neurons firing. A major sponge for the oxytocin and dopamine that get released during mating and "mate-approaching" behavior, this region likely encodes positive memories as well as the desire to repeat those memories. It also turns out to be one of the main areas of difference between monogamous prairie voles and their roguish meadow vole cousins. The meadow voles don't have many cell receptors for those neurotransmitters in that spot. Humans do, some more than others. In fMRI brain-imaging studies of humans suffering complicated grief, the nucleus accumbens is unusually active while looking at pictures of lost loved ones.

Basically, love boils down to this: a strong emotion attached to memories. Like prairie voles, meadow voles enjoy mating, but they don't seem to form the memories of a loved one because their brains aren't set up to receive the chemical signals to do so.

Even niftier than watching vole neurons, Donaldson can *make them fire*. Using a process called optogenetics, she sends a virus into the vole brains that inserts a gene from green algae. The gene turns on a protein in targeted cells that makes them receptive to light. Donaldson can deliver light to the brain through an implanted fiber-optic device, and now a sodium channel in those specific neurons will open, causing them to fire. By stimulating these neurons at the right time, she can, in effect, create an affectionate memory for those voles. It's the reverse of *Eternal Sunshine of the Spotless Mind*, in which mad scientists erase the exact spots in Jim Carrey's brain encoding his memories of his ex-girlfriend, Kate Winslet. When Donaldson lights up the special love neurons in the nucleus accumbens of meadow voles, they start acting like prairie voles. Donaldson can also manipulate their minds after they lose their partners to see whether jolting or silencing certain cells makes it easier or harder for them to accept the loss in the lever-pressing task.

So what does all this mean for us? Maybe nothing. Maybe voles are voles. But plenty of scientists seem to think our brains work in eerily similar ways.

"Can we modulate the memory in a way that might parallel or mirror rumination in humans?" she asked. The way we fire up our own memories matters in terms of our recovery. After a breakup, explained Donaldson, "we have to sort of reconfigure our entire lives. There's all these cues that remind us of someone we're in love with. Maybe we shared a house with them and we have pictures. We have to get to the point where we see those things and don't immediately break down, where it becomes almost bittersweet instead of debilitating."

YEARNING IS ONE of several core characteristics of grief. Others are stress and depression. Although it's tricky to compare these emotions across species, scientists have tried to study them in prairie voles who lose their partners (and yes, anthropomorphizing abounds). A former colleague of Donaldson's from the Young lab, Oliver Bosch, split up half of his vole couples in an experiment. After about five days, he subjected the males to various tribulations: he either dropped them into steep beakers of cool water (the so-called Forced Swim Test), suspended them by their tails, which were duct-taped to a skewer that was hung in a black box (aptly named the Tail Suspension Test), or placed them in a maze suspended high above the ground (the Elevated Plus-Maze). The latter creates a conflict situation: Will the voles indulge their exploratory nature by venturing into the exposed open corridors of the maze or stay in the closed corridors?

Compared to males who were still enjoying time with their mate, the partner-separated voles gave up flailing and fighting their way out of the swim beaker and the black box sooner. They essentially threw up their paws. Scientists call this listlessness "passive coping," and many believe it resembles depression. (It's worth noting that animal-rights advocates, who condemn such experiments as cruel, question the validity of techniques used to induce negative emotions, or even the application of the word *depression* to voles at all.) In the maze, the bereft voles spent less time venturing out into the open and more time acting stressed in the enclosed corridors (behavior likened to anxiety). These results were replicated by another experiment in a Chinese lab, which also found that newly single voles spent more time in dark boxes than exploring other rooms.

It wasn't only the voles' behavior that changed; so did their neurochemistry. The partner-separated voles produced more corticosterone, a stress hormone, than the voles just separated from their siblings, and their adrenal glands—which manufacture these hormones—also weighed more. Their high stress levels seemed to be driving their behaviors in the tests. When researchers shut down the voles' corticotropin-releasing factor (shortened to CRF; it's a major generator of the stress hormones), they fought and explored as vigorously as their happier brethren.

There's one more interesting finding worth mentioning. In all paired males, regardless of whether they were later split up or not, their brains made more of the stress-generating machinery, the CRF, than the never-paired. At first, Bosch and his team were surprised by this. Why would the brains of those in love rev up all that ammunition just to sit idly by? But then they learned that oxytocin from mating keeps the CRF in a quiet state. Unless, that is, the bonds of love are broken and the oxytocin molecules dry up.

Remarkably, it looks like the enhanced stress machines are there precisely *to respond to heartbreak*, even, in effect, to create it, Bosch believes. Because, as much as it hurts, the misery is intended to be adaptive. It drives us to reconnect with our lost partners after brief separations, and it keeps us coming back home. Here's the deal: pair-bonding, love, call it what you will, changes the brain. It changes it in some permanent ways that make us more sensitive to both joy and woe. It gives us a sense of something to lose.

Falling in love puts a loaded gun to our heads.

Unfortunately, what looks helpful for guiding us through short-term separation isn't so adaptive when your partner is actually gone for good. Now the stress doesn't easily resolve; it can just keep spinning in its relentless vacuum of oxytocin depletion. Interestingly, when the Bosch team gave supplemental oxytocin to the bereft voles, they behaved like normal happy voles in the stress tests. Perhaps this is why so many of us seek rebound sex.

In fact, Donaldson wants to try a similar remedy in the Heartbreak Hotel. If her bereaved voles mate again, do their brains start looking normal? If they don't mate, but start huddling with and grooming their friends and siblings, is that almost as good? Will they learn to press the lever less—*will they accept heartbreak faster*—if they are more social?

What about an even more anodyne solution? Donaldson believes that one day medicines may help alleviate heartbreak. Once used to treat tuberculosis, D-cycloserine, for example, has been shown to speed up certain kinds of learning in lab animals. "It's sort of like a smart drug," said Donaldson, who wants her voles to try it. "But it will only make you temporarily smarter." Taken at the right time—after, say, vole hubby fails to return home to Door Number One—it might help Mrs. Vole learn to move on. Someday, maybe, it could help people suffering from complicated grief as a short-term booster for therapy.

I found it astounding that so much carefully choreographed effort, money, and imagination are going into creating solutions like *a smart drug*. Take it and you have a magic window for transformative learning. Of all the possibilities for a temporary superpower like this—better understanding of climate change! cooperating for world peace!—the problem at least some scientists are rushing to solve is the one that actually scares us the most: losing love.

8

WELCOME TO THE EREMOCENE

Attachment

Suppose you threw a love affair and nobody came.
—LORRIE MOORE, *SELF-HELP*

I had no magic intranasal spray, smart pills, or incantations of any kind. What I had was a pile of bills and a parched lawn awaiting me back home, in a house that didn't feel the same. There is a Welsh word, *hiraeth*, which means a nostalgia for a place to which you cannot return. A new word, *solastalgia*, coined in the late Anthropocene, means a yearning for a landscape that no longer exists. Using the biggest heartbreak term of all, some people call our era the Eremocene, the age of loneliness, characterized by what the entomologist E. O. Wilson calls "the existential and material isolation that comes from having calamitously extinguished other forms of life on earth." I felt those losses harder now, a stateless exile from my former life. And, as the late writer Jim Harrison said, longing for our homeland can overwhelm us.

My strange, sunlit weeks at altitude had left me battle worn, twice dumped, and half a century old, but also filled with some sort of stubborn lifeblood. I did some laundry and charged the kids' phones, from which they had miraculously been separated for over a month. My kids! I couldn't wait to see them. I'd received a few sweet letters and caught rare glimpses of them as I stalked the camp's photo feed. They looked happy in the shots, covered with war paint, jumping into a lake, all framed by trees and vistas and blue skies. I had been relieved for them to experience some nature-filled joy and kid time, far away from the brittleness of their splintering home.

I drove through the Smoky Mountains to pick them up. My daughter clung to her friends for final-day hugs. In the car, the kids were full of tales of bees and downpours and coed dances. Back home, we eased into our real-world reentry together, watching some funny TV, cooking, playing with the dog, and buying school supplies. There were lots of moms and kids at the office-supply store. I would look at the women and think: Is she divorced? Is she? How about that one? They wore rings, all of them.

I wasn't just imagining that I was an outlier. Survey data from my DC neighborhood shows that 90 percent of the housing stock belongs to married couples. Although my husband had argued that half of marriages end in divorce (and therefore: stop overreacting!), I learned that among post-college-educated women who married in the 1990s, divorce rates are about 14 percent. Remarkably, the divorce rate has fallen by nearly half in my demographic since the 1970s. What's changed: economics (it's more expensive getting and being divorced), psychological reckoning (it's better known now that while children of divorce are not doomed, they do face increased odds of behavioral, academic, and future-relationship troubles), and the rise in helicopter parenting (see reason two). In addition, educated women, especially

college-educated women, are better able to achieve professional fulfillment and therefore personal fulfillment within marriage than they
used to be.

As with so many of life's adversities, your zip code and professional
status matter. For those with no high school degree, whose lives are more
likely to be challenged by economic precarity, social and physical problems, and structural racism, the divorce rate is still high—48 percent—
and for those with high school but no college degree, it stands at 38
percent. I knew I had more advantages than many people going through
relationship loss, and I felt more acutely aware of both what we shared
and what divided us. Suffering is not evenly distributed.

Every day I was grateful for friends and family. My son took and
passed his driver's test, and in those last days of summer we would drive
around the beltway to visit my 80-year-old father, who lives on the other
side of town. I went for early morning hikes with my friends, discovering
snakeskins and pulling old glass bottles out of the Potomac floodplain for
my friend Lauri's ever-growing back-porch art installation. Being with
my friends made me appreciate urban nature more. They weren't mountain snobs like me. Along the scruffy riverbanks, we picked paw-paw fruit
that smelled like unwashed sweatpants and made paw-paw ice cream,
paw-paw sorbet, paw-pawtinis.

I took my daughter to see *The King and I*, which I'd seen with my
mother when I was about the same age. The imperialist caricatures were
cringeworthy, but I could appreciate Anna the English widow's plight.
She disapproves of the polygamous Thai king's cavalier misogyny, and yet
she still loves him. Rather conveniently, he expires.

At intermission, I saw a text from Ennis. I hadn't heard from him in
about a week.

Thinking of you. How's the transition back to DC?

I told him where I was. *Your kind of musical*, I thumbed. *Man with
harem! Falls into unattainable love with writer/schoolteacher.*

Perfect, he replies. And an hour later, *Fucking unattainability.*

MY HUSBAND WAS gone, and Ennis was clamping about eight other nipples. So much yearning. So much unseemly, adolescent, self-involvement: daydreaming, journaling, wanting to lie around and listen deeply to music and tell all my friends exactly what was going on. It's tempting to dismiss the mistiness of this time as juvenile spaciness, but there was something to it. It was about the process of becoming. It was hard work. It demanded attention and space. Most adults think they can just be, but when the floor falls out, you realize you are no longer who you were; you have to plunge back in the jar like a pickled kipper and cure some more. You have to become again. It sometimes felt beautiful and special. My kids and I were all pickling at the same time. Maybe that's why we felt so tenderly attentive to each other.

I also felt plenty of anxiety as we settled into our fall routine. The word *anxiety*, or in German, *angst*, comes from the Greek word *ánkhō*, meaning "strangle," and the Latin words *angustia*, "tightness," and *angor*, "choking." The nineteenth-century German psychologist Emil Kraepelin described it as an inner tension that permeates both the physical and mental state. I felt it as a deep and restless fear of the unknown. Anxiety, argues psychologist Claire Bidwell Smith, is the missing stage of grief. Joan Didion describes it well. "The fear is not for what is lost," she writes in *Blue Nights* after the death of her daughter. "What is lost is already in the wall. What is lost is already behind the locked door. The fear is for what is still to be lost."

My disorientation was profound. My children are away at their father's for another seven days? I'm just cooking for one now? How will I afford health insurance? The other side of the bed is empty? How do I fix a broken lawn mower and *must I, really*? I drove around for months before realizing my car registration had expired. I couldn't keep up with the pace of change in so many large and small parts of my life. I basically stumbled around repeating one refrain inside my head: "Okay, *what*?"

(A confession and a digression: what I actually said was, *What the fuck?* I had taken to cursing like a sailor. Recently, psychologists at Keele University in the UK wondered if, since swearing is so common when we are under distress, there might be a useful purpose to it. Was Mark Twain's observation true, that "profanity provides a relief denied even to prayer"? To find out, they asked 67 undergraduates to list their top favorite cuss words, and then they made the students plunge their hands into a tub of ice water while either swearing or not swearing. Sure enough, the students who got to let their foul mouths rip perceived significantly less pain. This was true of both sexes, but especially true of the women, for whom cursing is considered the bigger transgression. The swearers were also able to keep their hands in the freezing water for about 30 percent longer. But why? The authors speculate that cursing—per battle and sports cultures—may amp up aggression, which in turn floods our emotional, limbic brain centers with some pain-numbing adrenaline, "downplaying feebleness in favour of a more pain-tolerant machismo." To which I mutter a hearty and offensive exclamation of approval.)

Every other week, the kids would leave, and the dog would go with them. I would hug them goodbye and scratch the dog's ears and my not-so-little son would drive with his brand-new driver's license in the old Subaru down the hill to the husband's house. Into the car went their suitcases, textbooks, dance bag, dog leash. The kids were growing closer to each other. I could see it. Maybe it was because my son, as a driver, had become useful and helpful to his sister, or because they welcomed the consistency they gave each other in a disruptive time, or because they were just growing up a little faster, together, into measured, practical people. They put on a game face for their parents, and for each other. It was impressive, even inspirational. It made me want to put on a game face too.

But alone, I cried. I would turn back to the drafty meagerness of my house. The dog hair and my children's skin cells floated in slow motion through the light streaming in from the windows, the last vestiges of life in a suddenly vacated space. One's children aren't supposed to pick up and leave their homes like this, week after week. While they were gone, I accepted nearly every speaking gig, whether for travel money or for real money. Yes, I'll go to your library in Akron, Ohio, and your middle school in Valparaiso, Indiana, in January. *Get me the fuck out of here.*

THE SHOCK OF heartbreak and the pain of loss are devastating, but what often comes after scared me more than anything. It is being—and fearing being—deeply, existentially alone. Loneliness, wrote Emily Dickinson, is "the Horror not to be surveyed." Yet surveyed it must be, if we are to reckon fully with its consequences and seek its dissolution.

Loneliness has been described as an experience, an emotion, and a primal terror. It is not always logical. Loneliness may be a normal and healthy response to feeling alone, but it easily becomes a pathology of self-awareness with respect to others because it is so often bound up with feelings of self-loathing.

In 1953, the psychiatrist Harry Stack Sullivan came up with a serviceable definition of loneliness: "the exceedingly unpleasant and driving experience connected with inadequate discharge of the need for human intimacy." At the core of loneliness lies a chasm between what you want and what you perceive you have. The British writer Olivia Laing describes her pain after a breakup in *The Lonely City*: "I want someone to want me. . . . It was the sensation of need that frightened me the most, as if I'd lifted the lid on an unappeasable abyss."

To understand why relationship loss hurts so much, it's helpful to start with the original bond, the one that sets us up for heartbreaks from that point forward.

Our lives begin with love. While loneliness may be a human concept, all mammals are social. We are named for mammary glands, which confer sustenance through the act of attachment, literally. For love to be a motivating emotion, you have to crave it. From an evolutionary perspective, all that necessary closeness, in turn, facilitated social learning, which led to play, communication, cooperative hunting, and complex offspring-rearing strategies, and for us humans, language and meaning and culture. You could say that it all started with milk.

Researchers have learned more about the power of attachment by watching what happens when it gets severed. Humans are the first mammals to undo—by accident and on purpose—our social networks and family bonds. As recently as the middle of the last century, many prominent Western scientists and parenting experts were arguing, quite persuasively, that children didn't need to be emotionally attached to parents or caregivers, and that it was in fact better if children in hospitals and orphanages were kept socially isolated to prevent the spread of germs. They warned parents in general, mothers especially, not to be too doting or physically affectionate with their newborns or children, lest the coddled youth become lazy, spoiled, and overly emotional. "When you are tempted to pet your child remember that mother love is a dangerous instrument," warned John Watson, a Johns Hopkins psychologist, best-selling author, and president of the American Psychological Association, in 1928.

Through the 1960s, many Western hospitals separated mothers and infants in the early days after birth. Most pediatricians (male, steeped in the science of the day) also discouraged breastfeeding in favor of formula, which was reassuringly quantifiable and uniform.

Despite the renewed attention to germ control, the orphans in foundling hospitals were still dying in droves. They were also, noted observers, listless, sad looking, and catatonic. As journalist Deborah Blum notes in her history of attachment science, *Love at Goon Park*, public sentiment began to shift slowly during World War II, when 700,000 children were separated from their parents in England and sent to live in safer country towns. These children did not thrive. Many became bed wetters, anxious, depressed, sickened by chronic infections. They were, in a word, heartbroken. "Nothing in psychology had predicted this," Blum writes.

The psych experiments of the mid-twentieth century, and still today, make one very glad not to have been born a lab animal. For understanding attachment, rats, mice, dogs, cats, monkeys, and scores of other creatures have been separated at birth, raised by surrogates, fed by machines, removed from mates, returned to mates, shocked, dunked, placed in isolation chambers, dissected, inspected, and implanted. But the insights gleaned from many of these studies have fundamentally altered the way we view (and now therapeutically treat) mammalian social drives, disorders, needs, and behaviors.

Two mid-century attachment researchers in particular stand out: John Bowlby, a British postwar psychiatrist, and Harry Harlow, a primatologist at the University of Wisconsin. It is hard to even talk about human attachment and its opposite—loneliness—without understanding their contributions. Born to a baronet in 1907, Bowlby was packed off to boarding school at a tender age. He understood the pain of forced separation. Later, as a young psychiatrist, he witnessed firsthand the downstream effects of having neglectful or abusive parents, which the Freudians tended to downplay as less significant than a patient's own neurotic sexual drives and other internalized imaginings. "There are few blows to the human spirit so great as the loss of someone near and dear," wrote Bowlby, who worked throughout his life to promote parent-infant attachment.

Partly inspired by Bowlby, Harlow's research employed rhesus macaques, who, like humans, are not just social but hypersocial. In his most famous—and infamous—experiments of the 1950s, Harlow and his colleagues placed newborn macaques in a confined space with either a cloth-covered monkey doll or with a metal-wire doll. In both cases, the babies grew up damaged, lacking social intelligence, unable to play or mate or function normally once they returned to their monkey groups. In one set of experiments, Harlow placed a newborn with both doll models, but only the wire mama held a bottle full of milk. Proving wrong the theorists who insisted infant "love" behavior was only a conditioned response to being fed, the baby monkeys gulped the milk as quickly as possible and then jumped back onto the soft mama, clutching her, stroking her, and nuzzling her. The babies were after more than just a meal. Moreover, they were able to bond far more to the cloth mother than to the wire one. But it was the real mothers who were best able to make the infants feel safe. Bowlby, Harlow, and their colleagues, like Mary Ainsworth, showed that securely attached infants have the best chance of navigating novel environments and future relationships.

Harlow, a workaholic and an alcoholic who lost one wife to divorce and another to cancer, understood loneliness. When his second wife, Peggy, grew mortally weakened by disease, Harlow suffered a severe depression. It drove him to seek replicating that state in his macaques with the hope of providing insights for treatment. But how best to make them depressed? It was John Bowlby who gave him an idea while touring the Wisconsin lab. There was one thing guaranteed to make the monkeys abjectly, certifiably miserable: social isolation.

Using a contraption he called "the pit of despair" in the late 1960s, Harlow placed single macaque youths in the bottom of a funnel from which they could neither escape nor see their parents and friends. The emotional results were immediate and long-lasting. "Depression," wrote Harlow, "results from social separation when the subject loses something of significance, has nothing with which to replace that loss, and is incapable of altering this predicament by its own actions." The depression, marked by lethargy, despondency, and an inability to interact with others, was very hard to treat once the animals came back to the social world. One thing that sometimes but not always helped was placing the scarred animals with young, clinging infants, starting with just a few minutes a day. Another thing that helped was placing them with dogs, suggesting that pets can be important in relieving loneliness and emotional trauma in humans as well.

In trying to study animal attachment, Harlow ended up reaching profound conclusions not only about love but about the loss of it. Love could, he understood, bring out the absolute best in us, but it could also cause our deepest suffering. Or as bell hooks baldly put it, "Romantic love is one of the most destructive ideals in the history of human thought."

LIKE THE VOLES who hadn't entirely given up yet, I was pinwheeling, restless, agitated, needing to move, getting on planes, pointing my face up to the light, trying to find beauty in the sky.

"Why am I incapable of sitting still?" I asked my therapist. "Shouldn't we examine this?"

She shrugged. "We could."

"Aren't I doing this wrong?" I asked. "Aren't I supposed to be like Pema Chodron and sit in my pain?" Chodron is the well-regarded Buddhist nun and author of *When Things Fall Apart*, which I'd been reading. "I'm supposed to be sitting in it, not keep running away from it?"

"You still seem to be feeling your pain," Julia said.

This was true. I was still crying every day. At least for a spell.

I went to California for some reporting and a friend's 50th birthday party. High in the Sierras, her mountain cabin overlooked a stand of firs and spectacular jagged mountains. As usual, the gathering was full of couples. But her mountain neighbor stopped by, a poet whose work I knew and admired. We strolled down the road to look at the moon. He was heavy lidded and handsome, like Mark Ruffalo with thinner hair.

He invited me to come back in a few days after my interviews. I was still trying to understand and express the intense carnal currents in my body. This man was asking me to share something. His poetry was quiet and quirky and beautiful. Here was yet another chance for openness knocking on my heart. I thought, *Let's give it a go, Ruffalo*.

He was enthusiastic, accommodating, playful, appreciative. All his parts worked. We spent the weekend in his room or hiking in the mountains. The pine-covered hills were glorious. I was bowled over by a kind of euphoria, a feeling that I was alive and connected to trees and men that were transpiring and redolent, if not exactly easy to access. Driving to trailheads, he complained that Faulkner overwrites and Hemingway says nothing. He was irascible about his editors, his ex-wife, and his last girlfriend. I knew he had a reputation for being difficult. He'd been fired from a university job, and he'd alienated colleagues at a magazine. I found myself thinking about his loneliness. He told me he hadn't slept with anyone in two years. He kept largely to himself. When he did venture into the nearby town for a beer, it didn't always go well. He'd gotten into two bar fights. In his basement, he kept six months' worth of supplies for the coming apocalypse. Even though he lived in the middle of nowhere, he thought it likely that his hideaway could be ground zero.

When it was time for me to leave, he hugged me and pressed a gift into my hand. I looked at it. It was an Opinel, the classic French wood-handled folding knife.

"I don't need this," I said.

"Take it anyway. You never know." Lonely people need protection, and I was a lonely person now too. I liked the feel of it.

GRANTED, I DID not know the poet well, but his actions made more sense to me months later, when I went to Utah to visit a young neuro-anatomist named Moriel Zelikowsky. Like Donaldson with her voles, Zelikowsky studies circuits in the brain in order to help people who are suffering. Instead of voles, she works with little black mice, and rather than studying heartbreak directly, she studies consequences of the social isolation it can create: violence and aggression. She landed here indirectly.

Zelikowsky had heard that socially isolated fruit flies become more aggressive when their brains release a signaling neuropeptide called tachykinin 2, or Tac2. She wondered if isolating mice from their cage mates might elicit similar effects. She parted the lab animals for two weeks, and then harvested their brains. By staining the tissue to glow green wherever Tac2 was present, she could consider the relationship among loneliness, behavior, and this neuropeptide.

"It was one of the more surprising discoveries I've ever made," said Zelikowsky, a married, energetic, sports-loving mother who describes herself as not at all socially isolated. She expected to see some lit-up regions, but "the whole brain to my naked eye looked neon green," she said. "There's such a massive upregulation of that neuropeptide across many, many regions in the brain," including areas no one suspected would release Tac2.

And the molecule, which had been literally bathing the brains of these lonely mice, seems to activate a long list of unsavory behaviors. When the mice have been socially isolated for a time, they become curmudgeons, and dangerous, too. Notably, this doesn't happen if they've been isolated for just a day or two, when they still act affectionate and even extra social (a day in a mouse life is equivalent to several months in a human life). But the longer-isolated mice will attack submissive intruders. They exhibit aggressive, anxiety-like behaviors consistent with a mouse version of paranoia.

"They show this increased fear and a persistence of fear," explained Zelikowsky, who has also inhibited the Tac2, and its effects, through a pharmaceutical that she believes could be a potential anti-loneliness drug for people. "If you are continuing to show fear responses when it's no longer appropriate, you're kind of limiting your other adaptive behaviors, such as finding mates, finding food, things like that. They spend a lot of time at the edges of an environment rather than in the center, similar to students who are nervous in a class sitting in the back row or hiding in the bathroom."

The mate-finding part interested me. I asked Zelikowsky to introduce me to the postdoc who is studying that piece of it. Tall and thin, he was wearing a woolen cap and holding a sandwich. This was Jay Love.

"Seriously, your last name is Love?" I asked.

Love told me he trained as an ornithologist, with a specialty in birdsong. Now he's applying what he knows about sound to mice. As he explained it, the longer-isolated mice find fewer mates than the social mice. For one thing, their mating calls sound flat. I couldn't hear the calls, because they are ultrasonic. But Love showed me a visualization of the sound by pulling up a bar graph of vocal range in different groups of the animals. Then he pulled up a series of scratches on a grid representing the range of syllables in the calls. Some of the scratched grids looked like Agnes Martin paintings, simple horizontal lines. Some looked more like complex Jackson Pollacks.

"I quantify each of these syllables and then compare them," he said. "The loner mice are not changing pitch as frequently."

I wanted to make sure I understood it. "So there are like the normal mouse guys going, *Heyyyy there, beautiful! Let me buy you a drink and we can make some magic!* while the loners are saying, *Um, can I maybe sniff your bum?*" I asked.

"Yeah. You could anthropomorphize like that."

When the loners try to mount a female, it's for a shorter time, and "they don't always achieve intromission," said Love. That, according to the dictionary, is science-speak for "the action or process of inserting the penis into the vagina in sexual intercourse."

Loners, aggression, paranoia. Was I crazy to compare the mice to the men I was dating, or to myself, for that matter? Not really, said Zelikowsky. "I think it does for sure translate to those feelings of loneliness one has that often accompany heartbreak or the end of a relationship."

Zelikowsky would likely not have used social isolation as an effective experimental stressor if it weren't for the foundational, controversial research of Harry Harlow. Eventually his work would inspire many, many other studies across species. As the decades passed, researchers have been able to examine neuroendocrine changes, brain-structure changes, and, most recently, gene-expression changes in lab animals and humans dealing with loss, loneliness, and isolation. The results have sobering implications, providing insights into how our brains and bodies cope with social pain. In addition to changing our behavior, how we feel around each other affects our physical health at a cellular level.

9

YOUR CELLS ARE LISTENING

A sad soul can kill you quicker,
far quicker than a germ.
—JOHN STEINBECK, *TRAVELS WITH CHARLEY*

That fall after the split, my physician was stumped about why my blood sugars were still so high, since I was fit and underweight. First she sent me to a nutritionist, who said my diet was good, so the problem must be stress. She told me to count four slow breaths before eating every meal, to meditate, and to eat slowly and mindfully. Then my doctor sent me to an endocrinologist, who ran some more tests and confirmed a surprising diagnosis: diabetes, type 1, or what's sometimes called type 1.5, which comes from the immune system attacking the pancreas. (In the more common type 2 diabetes, the pancreas typically still produces insulin, but the body's cells are resistant to it and unable to metabolize sugar.)

Type 1 is usually diagnosed in children. It's rare to be diagnosed as an adult, but when you are, it tends to progress more slowly. My glucose levels were stretching into the yellow as opposed to the red zone, but my insulin-making cells appeared to be faltering or, more precisely, slowly drowning. It's unclear exactly what damages these beta cells, but it appears to be driven by inflammation causing an enzyme to suffocate the cells in lipids. Since I still had some functional beta cells in my pancreas, I was told I might be able to delay the need for daily insulin shots if I could manage the carbohydrate load in my diet, work on my stress levels, and exercise after eating.

I now had not one but two autoimmune diseases. I already had a thyroid disorder called Hashimoto's, which I'd been ably managing for years. Diabetes was a whole different beast, capable of killing you while you sleep and maiming you in all sorts of medieval ways (the eyes, the ankles, the arteries) over time. I would be a fool to ignore it, yet I dreaded obsessing about another scary thing, especially when the very fact of stressing about scary things was contributing to the disease. People with diabetes say it is like a part-time job. It demands your attention, your vigilance, your brain's executive network all day long.

WE'VE ALL HEARD that stress influences our immune systems. When we are under threat, neurotransmitters like norepinephrine cause our bodies to direct resources where they are needed. Our heart rate, blood pressure, and respiration increase while things like digestion, fertility, and ability to fight diseases fall off. When stress levels stay high over time, though, the neurological signals prolong and amplify inflammation, and that in turn makes us susceptible to a steady rain of long-term health woes. As an evolutionary adaptation, inflammation can be useful, even as a response to sustained stress and depression. Before 10,000 years ago, one of the biggest threats to human life was infections from injuries; inflammation helps fight bacteria in wounds. Even today, in high-pathogen areas like those around the equator, whole-body inflammation is likely to decrease the risk of death. But in low-pathogen areas in higher-income countries, persistent inflammation increases the risk of death from chronic diseases like cancer, heart disease, Alzheimer's, diabetes, and others.

An old college friend, weirdly, had also become diagnosed with adult-onset type 1 diabetes some months after her marriage exploded. Now she wore an insulin pump 24 hours a day and worked out on a stationary bike after eating carb-heavy meals. There was no diabetes in my family, nor in hers. Was divorce diabetes a thing? It is impossible to know the answer to that, at least as far as she and I are concerned as individuals. What is known is that autoimmune diseases can appear after a stress "trigger," according to Stanford University's Michael Snyder, a molecular geneticist and himself a midlife diabetic.

"Diabetes is known to be associated with stress," he said, explaining that under certain circumstances, our DNA can start expressing diabetes promoter genes. High levels of cortisol, known to be elevated during emotionally stressful times, interfere with the production and regulation of insulin. Someday, medicine will be sophisticated enough to watch this happening to people in real time. A passionate evangelist of personalized medicine, Snyder himself walks around wearing multiple biosensors, for example in his watch, that monitor everything from heart rate to blood glucose to oxygenation to body temperature to sleep quality.

While scientists have known for decades that death and disease increase (substantially!) after divorce, some are now trying to investigate which antibodies, inflammation markers, and gene sequences can lead to trouble. Researchers at Ohio State University found that adults who were struggling emotionally with their recent divorces (in this study, the time frame was within two years) produced fewer natural killer cells, which are important for fighting cancer and other diseases. They were also more likely to acquire viruses like Epstein-Barr than their married peers.

As psychologist David Sbarra, a professor at University of Arizona who authored a sobering divorce-and-health review paper, told me, "There is an inflammation story related to divorce."

———

I DIDN'T KNOW just how much that story would change the way I think about love, loss, and human health. It struck Steve Cole the same way.

In the mid-1990s, Cole was a young genomics researcher more interested in viruses than in social relationships. Then he joined a team of epidemiologists and psychologists wondering why some gay men with the HIV virus were getting sicker and dying at a faster rate than others. Cole, who works in psychiatry and biobehavioral sciences at the UCLA School of Medicine, looked at a group of 72 men over an eight-year period, comparing a number of psychological traits with counts of their immune systems' T cells, as well as how long it took them to get symptoms of disease. The strongest factors determining the speed to AIDS and death, they found, were sensitivity to social rejection and whether or not the patients were public about their sexuality. Closeted men, who faced the stresses of secrecy and possible discovery on top of the stress of their HIV status, got sick two to three years earlier than men with HIV who were out. Shy men, who constantly worried about how others perceived them, experienced similar declines. Both groups had nervous systems that were easily triggered into a stress state by social circumstances. Cole found these men produced more of the hormone norepinephrine—a key driver of our fight-or-flight response. The hormone made T cells more vulnerable to attack by HIV, with the virus replicating 10 times faster than it would otherwise.

People with a terminal, brutal, and stigmatized disease no doubt faced an unusual amount of stress. So Cole was surprised when he was approached in the early 2000s by a psychologist at the University of Chicago named John Cacioppo asking him to expand his genomics work into a far more common and mundane population: lonely people. Cole worried that there would be little in the biology to work with, that loneliness "was just whiney, complainy stuff that happened in the brain." But he agreed to analyze some blood. Cacioppo, he knew, had played a key role in popularizing the concept that people who self-identified as feeling lonely—as well as people who were literally socially isolated—were known to suffer from more diseases and early death than those with strong social-support networks.

Moreover, loneliness had all the hallmarks of an epidemic in America. It was widespread, growing, and costly. Cacioppo was dismayed by the prevalence of loneliness—about 20 to 25 percent of the population, according to deep databases like the decades-long Health and Retirement Study—and by its impact. Chronic loneliness increases the risk of early death by 26 percent, similar to being obese or smoking. But why? Cacioppo wanted to know if there were unique cellular and genetic markers in lonely people that might yield clues to the pathways to disease.

Along with Cacioppo, Cole and four other colleagues published a remarkable paper in 2007. It was the first to consider the effect of social factors on gene expression in the immune system. What the team found in analyzing blood samples from a small group of relatively healthy adults was that social connectedness—how emotionally and socially connected people feel to others—altered the activity of two key sets of genes in white blood cells. These in turn wrought wildly divergent immune responses, particularly with respect to viruses, bacteria, and inflammation. Cole was stunned to find such clear molecular signatures of our psychosocial state. He called the results "beautiful genomics," as well as a call to arms. "As I thought about it more," he told me, "I realized this is really bad. This is basically a molecular recipe for early death."

After analyzing many other data sets, Cole now calls loneliness one of the most toxic risk factors known to human health. He's been pioneering the field of social genomics ever since. Cole takes the epidemiological work of people like Utah's Bert Uchino, who reveal the health slides of heartbreak, and explains why and how they occur.

When I first phoned him while I was driving in the Colorado flats between the Denver airport and Boulder, it became apparent that Cole was the rare academic who is both extremely accomplished in his field and also compassionate, curious, and generous enough to offer up a substantial amount of time and expertise to a journalist. I caught him in a reflective state; the scholar who pivoted the trajectory of Cole's own career, John Cacioppo, had just died from cancer at age 66 the day before.

"Intellectually, I feel a lot lonelier today," he said. "John was an esteemed psychologist studying the technical side of cognitive processes, and one day he said, 'I'm going to study loneliness for the rest of my life.' I don't know why, and now I never will," Cole said. Cacioppo wasn't lonely. He was in a happy late-in-life marriage to a colleague and collaborator, Stephanie Cacioppo, a neuroscientist who specializes in studying the bonds of love and affection (and the loss of those bonds). A few months before he died, John Cacioppo told the *New York Times*, "One of the secrets to a good relationship is being attracted to someone out of choice rather than out of need. We weren't running from anything aversive. We were moving toward something that was really unique."

Cole told me that Stephanie had written to him, saying she would now have to put into practice the many loneliness-buffering interventions they had studied together.

I pulled over, parking on a gravel shoulder, typing madly into my laptop as tumbleweeds blew across the hood of my car. We had a wide-ranging conversation about grief and divorce. I told him about my diabetes diagnosis and how it made me curious about the connection between our emotions and our immune systems. It didn't seem helpful or adaptive that our bodies fall apart at the same time our social worlds crash down, but Cole made me think differently about this and many other things over the course of what would become a two-plus-year-long dialogue.

It shouldn't be surprising that our immune systems become implicated when we are emotionally crushed, he said. Still, we don't expect it. We think the damage is all in our heads. "We think of relationship loss and isolation as pragmatic problems," he said, "because we are overly cognitive beings." We are fairly good at trying to solve problems like how to cook a meal for one person and how to transfer auto insurance. And yet that cognitive mode soon becomes insufficient. "Our bodies want what they want, warmth and the feeling of being understood by a partner, and now it's not there. Shock and panic set in."

Even though heartbreak is nearly universally experienced to greater and lesser degrees, we still don't take it seriously enough, he said. And unlike Cole, we don't all have the ability to witness the unexpected cellular carnage taking place.

"This is one of the hidden land mines of human existence," he said. And this was coming from a man who had spent years studying AIDS and cancer. If you can't get through heartbreak—if it continues to pummel your self-esteem and ability to interact meaningfully with others—you're in trouble. Being a functioning, relational person "depends on morale and enthusiasm and sparkle, and if you can't muster that, we now know, it's a death spiral."

Death spiral. The words reverberated through my rental car.

"What the hell are we supposed to do?" I asked.

"Don't be heartbroken forever."

By the end of the call, he'd given me a helpful pep talk about doing things I loved and staying attuned to the grounding presence of my kids. He also extended an invitation to stop by UCLA to give him some of my own blood. We'd take a look inside my immune system. Specifically, he would peer into my monocytes, which are white blood cells. Unlikely as it sounds, they listen for loneliness.

I hung up and looked out over the featureless prairie. To my left stretched the Great Plains. To my right, beyond the passenger mirror, rose the Rocky Mountains.

More than ever, I felt the urgency to follow Paula Williams's advice: I loved being in the wilderness, where I knew I could find some awe and maybe jump-start the process of calming the fuck down.

As it happened, an opportunity would soon come my way.

10

THE BODY DOESN'T LIE

To aim for the highest point is not
the only way to climb a mountain.
—NAN SHEPHERD, *THE LIVING MOUNTAIN*

It was early fall in DC. I weighed 106 pounds.

I had sobering meetings with my divorce lawyer and with a financial planner who presented alarming columns of numbers. My right eye kept twitching and I sometimes got heart palpitations. My laptop wasn't working well. One key—the F key, notably—wouldn't work no matter how hard I stabbed it. Even my computer was trying to erase me.

This wasn't easy on my husband either. His back seized up, and he was worried about an upcoming marathon. He might DNF (did not finish), a stain for a competitive runner used to topping out in his age group. I realized that's what we did to our marriage: DNF.

The warnings of social genomics whiz Steve Cole kept echoing through my head. *Don't be heartbroken forever.* I visited an acupuncturist named Bernie. I lay faceup on his table. To assess me, he placed two fingers from his left hand on the top of my left foot and two fingers from his right hand on my collarbone, hovering above me like an eagle.

"Are you in shock?" he asked.

"Um, that is possible." It was a discouraging admission. Months had gone by since my husband had left. How long can shock possibly last?

"The body doesn't lie." Bernie sunk needles into my stomach. He arranged tufts of mugwort atop the needles so they looked like Lorax trees. Then he lit the mugwort with a match. I felt like I was lying on a funeral pyre. The needles grew hot and smoky. I found it visually arresting and oddly relaxing to be half dead and on fire.

PSYCHOLOGISTS SOMETIMES DESCRIBE heartbreak as an attachment injury, especially when it involves rejection or betrayal. People say French is the language of love, but it may also be the language of heartbreak. The root of the word *betrayal* comes from the French verb *trair*, "to hand over," as in handing someone over to the enemy, similar to *traitor*. From that we got the Old English word *bitrayen*, "to mislead or deceive." The *be* root means "thoroughly." The *Cambridge English Dictionary* defines *betrayal* as "the act of not being loyal when other people believe you are loyal." And the *Oxford English Dictionary* adds "an abandonment of something committed to one's charge." Betrayal is a particular kind of uprooting of trust, an undoing of the truths you once believed that may lie at the core of your social identity and a fraying of a connection once held near inviolable. Feeling betrayed drives acts of desperation, literary feats of passion, and some of the most moving passages of theatrical tragedy, from the screams of Medusa to the Sopranos. *Et tu, Brute?*

Psychologists Saul Miller and Jon Maner parsed how men and women differ in their responses to romantic betrayal. Men tend to get angry, sometimes violent. This, they say, may be a vestige of evolutionary urges to reclaim lost social status by fighting off competing males who are after their mates. Women respond to betrayal with anger, sadness, and "coalition-building," among other things. One study found that wives are six times more likely to experience major depression after discovering their husband's infidelity than women who did not suffer what the authors called "a humiliating marital event." Another study found that, compared to men, women report more self-destructive behaviors following infidelity by their partners, including consuming drugs and alcohol and having unprotected sex. The longer the partnership, the worse they reported their mental health.

What about emotional betrayal versus sexual betrayal? It's complicated. A number of studies have shown that in heterosexual relationships men get more upset if their mates betray them sexually, while women are more distressed by emotional betrayals. Classic evolutionary theories of jealousy posit that a man has more to lose if his partner births a child who does not carry his genes. For women, on the other hand, emotional loyalty is necessary if their mates are going to stick around and share resources. But a couple of recent studies suggest it's not quite this simple. In both men and women, the degree of distress depends on an individual's style of relating. If the betrayed partners have a so-called "secure" style of relating, in which they feel self-confident and can easily maintain intimate partnerships, both sexes are then more likely to be upset by emotional infidelity.

Either way, for most of us, infidelity hurts. There's some debate in the scientific literature about whether it feels worse to be left for another person—a "mate poacher," as they're sometimes called—or left for some other, perhaps unexplained reason. A 2017 study at Cornell University asked 200 participants to imagine receiving a text from a romantic partner breaking off the relationship with no explanation. Researchers then asked them to imagine hearing from a mutual friend that the lost lover either didn't want to date anyone right now or had decided to date someone else. Not surprisingly, the participants reported feeling worse, and for longer, if there was (even an imaginary!) someone else. The other finding was that right away people tended to assume there was a competing love interest, probably because we are hyperaware of our relative strengths and weaknesses. "Most of us do not enjoy feeling that we are being directly compared to someone else," said study author Sebastian Deri. "We're attuned to the subtleties of status and it feels bad to be diminished."

Infidelity has probably always stung, but it may feel worse than ever before, because so many of us now marry for love. As psychologist and relationship expert Esther Perel puts it, "When people marry their soul mates, infidelity is traumatic."

It feels so bad that we are likely to imbue emotional and sexual infidelities with more weight than they might deserve. In a Gallup poll of over 1,000 Americans, 62 percent said they would leave their spouse after discovering an infidelity. Partners stray in an estimated 20–40 percent of heterosexual marriages. In about a quarter of divorces, infidelity is cited as a major reason for the split, but tellingly, spouses disagree on its import depending on who is doing the straying. It's easy (for the betrayed) to make betrayal the central story of a broken marriage, but of course it's rarely that simple.

Not everyone will experience heartbreak as trauma. Emotional trauma, according to the Substance Abuse and Mental Health Services Administration, "results from an event, series of events, or set of circumstances that is experienced by an individual as physically or emotionally harmful or threatening and that has lasting adverse effects on the individual's functioning and physical, social, emotional, or spiritual well-being." Like loneliness, trauma is a subjective experience. And like grief, it has the potential to linger. Among those whose hearts break, most will recover before the psychological wounds turn into post-traumatic stress.

If trauma was what I was experiencing, it made sense to look into how people deal with it. I didn't have PTSD—not enough time had elapsed— I just had the *T*. In October, when my editor at *Outside* magazine asked me if I wanted to report a story about an unusual group of backpackers, I said yes. They were women from Atlanta, and they were sex-trafficking survivors. It sounded like an intense, important story and I was ready to get out of my own head.

I LANDED IN Denver and met Elise Knicely at an airport café, where I immediately attacked a low-carb guacamole dip. We had some time before the women would arrive. Elise, the founder of She Is ABLE, the nonprofit sponsoring this trip, was 27 years old and beautiful, but she looked worn-out. The organization was only a year old, understaffed, and underfunded. In that year, it had put together 16 trips taking 100 sex-trafficked women on one- to four-day wilderness trips. And then there was the emotionally draining nature of the work itself.

Why did Elise think a few hours or days in the woods could possibly help course-correct a near lifetime of abuse and exploitation, often compounded by addiction? At first, it was just intuition layered with sorority-girl optimism. An Alpha Chi Omega from the University of Georgia, she played a lot of sports and enjoyed the outdoors as a girl. After college, she tried corporate consulting, but hated it. A solo trip around the world landed her in India, where she shadowed an organization fighting sex slavery in the slums of Mumbai. She met women forced to live in rooms that looked like cages. "The encounters changed my life," she told me. "I had a moment in India where I was like, man, this is just it for me." She came back to Atlanta and took some of the girls from a shelter hiking.

Then she started reading the research. Increasingly, there's evidence that time in nature can help build strengths these women particularly need: self-regard, peace from the anxiety and hypervigilance associated with trauma, a healthier and more connected relationship with their bodies, and an environment that strengthens social bonds and support. The bonding piece made me think of the work of Utah mouse researcher Moriel Zelikowsky, who found the only thing that helped her mice—traumatized from loneliness—was hanging out with other mice.

Elise looked at me across the Formica table. "A voice said to me, E, this is what you're made to do."

WE FOUND THE women in baggage claim, six of them, varying in race and age from 20 to 37, carrying pillows and wearing sweatshirts. Only one, Sarah (all the participants' names have been changed), had been to Colorado before, on a similar program that brought women from a shelter to a guest ranch for a few days. She had graduated from being a resident to being an administrator in a group rehabilitation home outside of Atlanta. The women came from two shelters. Embarking on a four-day backpacking trip in the mountains for them would be part reward, part recovery. They were all nearing the end of yearlong treatment programs for substance abuse, sexual abuse, and other problems related to being trafficked.

Like many people, I had some misconceptions about them. When I told friends I was working on this story, they asked what countries the women were from. But the vast majority of sex-trafficked women (and a surprising number of young men) in the US are American, representing a range of backgrounds. Many come from middle-class families and suburbs. Technically, the term *sex trafficking* applies under three conditions: force, fraud, and coercion, according to Julie Laser-Maira, a trauma psychologist and trafficking expert at the University of Denver. Atlanta, with an underground sex economy that is estimated to generate $290 million per year, is one of the country's top trafficking cities. Cases in Georgia rose 41 percent between 2014 and 2015, according to the National Human Trafficking Resource Center. The trade thrives through illicit websites. Many victims are minors; Atlanta tallies hundreds of underage trafficked girls per month. They can have a dozen or more clients per night.

The fact that the women clutching their duffel bags and offering shy smiles looked familiar was both disarming and unsettling.

From the airport we drove in a rented van to a warehouse in West Denver, the headquarters for the group that would be guiding and outfitting us, Expedition Backcountry Adventures. "I can't believe we're in Colorado, y'all!" squealed Sarah. We cooked burgers in the cavernous space. Sitting around a large plywood table, the women recited tales of broken hearts and broken promises: their own children taken away by state authorities, neglect or abuse from their parents, pimps and relatives who refused to bail them out of jail, husbands or boyfriends killed or imprisoned. All had been betrayed by people they trusted.

Asta, a 33-year-old with long straight brown hair, had a rough start in life. Her mother was mentally ill and her father, she said, was a sexual predator. She was adopted by her grandparents when she was nearly five. The rest of her childhood was pretty typical: suburbs, school, church. She had a daughter when she was 20 and a son six years later, but the father of her children ended up getting deported. After that, she became addicted to crack and alcohol. She lost custody of her children.

Tamara, a quiet 30-year-old with a small tattoo at the base of her thumb, talked about losing custody of her six-year-old son, who now lives with his grandparents. She began to cry.

"Can I pray for you?" asked Kris, a broad-shouldered blonde. Tamara nodded. Kris offered her hand.

"Lord, please help heal every crevice of her broken heart. Help renew her mind during her absence from her son. Meet her where she is, whether it's in a warehouse or on a mountaintop. I know, God, you have a purpose for her life." Tamara sobbed harder, then Kris started to cry, and pretty soon we were all sniffling amid the camp stoves and Nature Valley granola bars.

This is pretty much how it would go—from group laugh to group cry to fervent prayer—during the next few days. And there was something about that intense cycle that seemed to help move difficult emotions along. After months in rehab, the women were used to processing and group therapy. They were so good at it that the trip leaders didn't always feel the need to step in.

"I've learned to let the women do most of the talking," Elise told me.

We unfurled our rented sleeping bags and pads on the concrete floor of the warehouse beneath stacks of dried food and first-aid kits. The ceiling loomed high above us. I awoke in the middle of the night unable to sleep. If I had been feeling miserable before the trip, now, after hearing some of their stories, I was feeling something else: anger. On that warehouse floor I had glimpsed a portal into the primal rage of women who are mistreated by men. And indeed, that portal was about to crack wide open around the country that very day, although I didn't know it yet. This was the day the story would drop about allegations against Hollywood executive Harvey Weinstein. Two weeks later, the #MeToo hashtag went viral on Twitter after being introduced by activist Tarana Burke some years earlier.

But for that night, it was a lonely reckoning. I got up in the dark and crawled into an empty office, where I fumbled to record some thoughts into my voice recorder. I talked about how these women have led incredibly hard lives. I was aware of my own privilege but also of the thin margins separating the lucky from the unlucky.

I still had a house, at least for now. I had my kids, half-time. I was not battling a substance addiction. I was curious to know if these upcoming days in the wilderness would help the women gain some sense of self-worth and peace, and if the trip would help me get there too.

AT DAWN, I rejoined the others to eat some wan pancakes and pack up our gear. Being with the group immediately cheered me up. They were bravely lacing up their hiking boots and trying on their new donated fleece jackets and brightly colored knit hats. "Girl, you look good in yellow," said Rochelle.

Elise appeared as the polar opposite of her charges. Tall and wispy, regally poised, with crisp outdoor gear and a confident stride.

"I like your cap," Kris said to her. "What does that mean, *Patagonia*?"

"Oh," said Elise, smiling, "it's an outdoor brand. I used to work in one of their stores."

Asta seemed quiet. I asked her how she was doing. She told me she had spent the night in her own kind of echo chamber, replaying bear growls in her head that she had learned on the internet. She wanted to be ready. Although she'd spent plenty of nights outside, usually under a bridge, she'd never spent a night in the mountains.

"I am nervous," she said. She spoke in a southern accent. She wore leggings and a loose sweatshirt, her long brown hair hanging straight. "I was homeless for a period of time, so the streets is what I'm used to. I was outside for three years in Atlanta. The mountains I'm quite intimidated by."

She mentioned bears and cheetahs. I reassured her there are no chee-tahs in the Rocky Mountains, but it was small comfort.

"Are you really more afraid of bears than men?" I asked.

"I know how to survive back there. This is unknown territory."

Fears can be outsize when you've spent much of your life afraid.

She met her first pimp when she was 26 and addicted to drugs. He advertised her services in a number of southern states through Back Page, a now-defunct website for escort services.

"Your ad is put on there," she explained. "And men can call and request you and then they come to your hotel room wherever you are and you have in-calls and out-calls." The crack she was on meant "decision-making wasn't really an option. My choices were made for me, but I allowed that door to open and then I couldn't close it." Even-tually Asta was found living under a bridge by a disguised rescue van that took her to a Christian ministry. After two weeks of near-coma-inducing detox, she was transferred to a residential facility called Freedom Hill.

Getting in that unmarked rescue van, riding out withdrawal: that was bravery. Still, coming to Colorado required an unusual leap of faith.

"I'm a runner," said Asta, not referring to athleticism. "I don't welcome or receive. I push away. I'm always expecting the worst."

Chelsea Van Essen, an effervescent, curly-haired guide, understands pervasive fear. The 26-year-old sat Asta and the others down in a circle before we loaded into a transport van. "One thing we want to just focus on as a theme of the trip is safety. Every day we will let you know what's coming," said Chelsea, a recent graduate in trauma psychology from the University of Denver. This was her first trip as leader for this outfitter, and it was the outfitter's first trip working with a group of women like this. Together, they would be challenging the orthodoxies of the wilderness curriculum.

"It has to be understood from the get-go that this is not going to look like a quote unquote normal trip," she told me. In some ways, it would look like less-than: fewer miles; fewer vistas; fewer scalable cliffs, rapids run, and canyons plumbed. But ultimately, its goal was to combine what Chelsea knew about adventure with what she also knew about the way the body holds trauma. "It's obvious the women here have done a ton of healing before this trip and they will do a ton of healing afterward, but this should be like a burst of catalytic energy."

THE WAY EXPERTS think about emotional trauma has slowly but dramatically changed in recent years. Freud believed that childhood sexual abuse was often the fantasy of an immature mind. But he was closer to the mark when he said, of a man he saw in 1895, that he was "suffering from memories."

Ideally, memories should reflect the past in a way that rationally informs the present. Almost got bitten by a scorpion once? Now you empty out your shoes in the morning. Fear memories are more strongly encoded (as are emotional memories in general). You will really remember the shoe routine. But sometimes the memory-coding metastasizes. Trauma is increasingly recognized as a memory disorder, in which searing life events continue to act upon the mind and body of the sufferer as though the events are still occurring. Traumatic experiences can leave deep and active traces in the parts of our brain that process fear and sensory perception.

Traumatized brains are fundamentally different: they show heightened reactivity in the amygdala, which reads threats, and less activation in the insula, which transmits signals to and from our bodily organs, joints, and muscles. Many trauma therapists like Chelsea believe that physical movement can help reignite the dimmed circuitry between bodily sensations and the brain. This, in turn, can help trauma sufferers stay rooted in the present. If you're busy feeling what is happening right now, it's harder to continue feeling the traumatic events of the past.

These ideas were new to me, and in some ways counterintuitive. How can the brain not be good at sensing inputs in the present moment? But the more I thought about it, the more it made sense. I'd seen it in others, for example in the hypervigilance or extreme impassivity of some veterans I'd spent time with on past wilderness trips, and I wondered if I could see it to some degree in myself. My digestion, my sleep patterns, my twitching zombie state were likely persistent memory hangovers from a combo of psychic blows delivered six months earlier in the split, plus maybe even childhood trauma I didn't often consciously think about. Was my nervous system still reacting to those blows? Is that why it was stubbornly residing in the hard, unyielding country of heartbreak?

WE PILED INTO the roomy white van. First stop: rock climbing.

"The van is new and needs a name!" said Chelsea. "What shall we call it?"

"Betty White!" yelled Tamara.

We agreed.

Betty White's first stop was Clear Creek Canyon, which reaches up into the Front Range from Golden. It was a cold October morning, and the women huddled behind the van to pull on more layers and lace up their climbing shoes. At a crag called East Colfax, named after a boulevard in Denver once known for street drugs and prostitution, we met up with our climbing guide, Aleya Littleton. Littleton is not your typical rope junkie. Diminutive, energetic, and infinitely patient, she's an adventure therapist who specializes in sexual trauma. She had already fixed ropes on three short easy routes, and she helped the women into their harnesses.

"I'm just going to watch," said Kris, gazing up the climb.

"Me too!" said Asta.

"It's your choice," said Littleton. But soon she made it irresistible. "Listen to your body," she coached. "You're going to be going from living in a horizontal world to a vertical one, so use your muscles in a way you don't usually use them, settle down into your legs, shift your weight. I notice if I breathe and find my center I can zoom out, open my focus. If I'm anxious, that focus closes down."

With Littleton belaying her, Tamara practically jumped onto the wall. She was a natural.

An hour later, everyone had climbed, some twice. There was much hugging and high-fiving.

Littleton gathered the group in a semicircle. The sun had entered the canyon, a breeze blew upstream. The women blinked happily in the light, like they couldn't quite believe they were here.

"What was that like for you?" asked Littleton.

Rochelle, a 33-year-old with piercings in her nostrils and upper lip, spoke first. "I noticed once I started trusting, it got a lot better. What are the odds I'm going to fall down the side of a mountain? Pretty low. When you said, 'Take little steps,' that really ministered to me. Baby steps. Baby steps."

Asta nodded. "I'm the only one who tells me that I can't do something. I tell myself that too much."

Littleton looked empathetic and also pleased. The activity had worked as hoped. Later, gathering the harnesses, she told me about the monthslong programs she leads. "It's so phenomenal to watch women go from anxious, with no voice, unsure, ungrounded, to total badassery on the rock," she said.

Climbing as metaphor may seem obvious. You have to trust people holding your rope. You have to find your breath and live in the moment. You move one step at a time while also looking ahead. You pull yourself up and cheer each other on. But the banalities of the metaphors don't make them less profound, and the benefits—from both the mental and physical effort—reach into unexpected places.

As Littleton explained it, healing trauma is complicated. Emotions reside in parts of the brain not related to linear thinking. Simply talking about traumatic memories doesn't work. Healing involves both separating fearful emotions from the bad memories and also bringing the nervous system back to the present. She advocates combining talk therapy with nonverbal practices like adventure sports and mindfulness. The Georgians have taken more risks than they expected while in their harnesses. But trust is an unfamiliar feeling. They don't trust others, and they don't trust themselves. They'd been heartbroken by fractured bonds of affection, and also let down by their own actions. Being unable to trust yourself is perhaps the biggest heartbreak of all. Everyone here had experienced relapse at least once. "It's always a step away," Asta told me.

The Georgians were helping me understand the urgency to heal trauma. To do so would mean transforming not just themselves but also their families and their communities.

BY THE TIME we arrived at the Monarch Lake trailhead east of the town of Granby, the sky was dumping graupel, a combination of rain, snow, and sleet largely unfamiliar to this group. It took us over two hours to hike the three miles to our camp, crossing into the Indian Peaks Wilderness. On the edge of some patchy meadows I spotted a mother moose and her baby, and showed the others, who watched wide eyed and smiling. A mile higher, white patches of snow glowed like spotlights on the forest floor.

"Well, this is not what I was expecting," said Kris. "It was 90 degrees in Atlanta." That night, after some singing by the fire, we heated up water to pour into our water bottles, which we tucked inside our sleeping bags. We wrapped our bodies around the warmth and tried to sleep.

The next day, thankfully, the clouds cleared, the temperature warmed to the 50s, and the gifts of nature arrived on cue. We smelled the vanilla bark of a ponderosa pine, walked across a log bridge over a fast-flowing creek, and picked our way over slippery roots and stones. Hiking, someone started singing a girl-power song from a Disney film. It reminded me that these women were still yearning for a happy childhood they never had. They talked about times with their kids, their favorite movies. For some moments that afternoon, it felt like innocence, youth, and even joy could all be reclaimed.

Eating supermarket bagels and canned salmon, I sat with Rochelle in a sunny spot. She had an easy laugh that belied her difficult past. I asked her how she was feeling. She smiled and groaned at the same time.

"Exhausted," she said. "But I am honestly doing a lot of self-reflection. Everybody else was hiking so fast, and my chest was hurting and my lungs were hurting and my shoulders were hurting and my knees hurt. I just have to keep on being realistic that, you know, I didn't take care of my body over the years and I have smoked drugs for a long period of time."

Rochelle's sexual abuse started when she was five, by friends of the family and by relatives. A ninth-grade dropout, she got sent to jail for the first time in her 20s for cashing a fraudulent check for a friend. "I didn't really have an identity," said Rochelle, who is now 32 and has dark kinky hair and bright eyes. "I was only good as long as I was being used. I did not know how to say no to people. I didn't really care about myself." That's how getting pulled into trafficking starts, she said, and also why it's so hard to get out.

These women fell victim to their pimps because they'd internalized the messages that they deserved it on some level. They believed that their worth was tied to their ability to be useful to men.

To live under these conditions, and to survive them, many abuse victims go largely numb. "I came to a place where I had to be okay with being raped," Rochelle told me. "I had to be okay with it because it happened every single day for a really, really long time. Out of automatic defense mechanisms or just trying to keep myself safe, I had to disconnect physically."

In no way could I compare what I'd been through with what these women had been through, and yet here's what I found myself unexpectedly identifying with: their bottomed-out self-esteem, their sense of not being in control of life events, their struggles to regain some self-determination. These thoughts unsettled me, but also filled me with admiration and compassion. I couldn't believe how brave Rochelle and the others were, how far they'd come. This was the contagious part. Not the trauma—although it is everywhere—but the survival instinct. We were all feeling it that day in the sunny woods.

I WAS CURIOUS about the numbness and dissociation Rochelle had described. Later I talked to Denise Mitten, a professor emeritus in adventure education at Prescott College. She told me that being in nature is particularly helpful with this aspect of trauma. In a dissociated state, people experience their minds and bodies as separate. Often, they try not to feel their bodies at all.

"If I've disassociated and I get to an environment that feels welcoming in any way, my body will start to relax. The kind of nature is very important. People want calm nature and now there's research that shows coherent nature is helpful too, not a wild tangled mess."

I asked her why it was so important to stop the disassociation.

"If people are disassociated, they lack self care, they're more inclined to do things that not healthy because they don't feel it. You want to do it at a pace that works. I have a zillion metaphors for how being in an environment can help us associate in our brains to help people feel connected and to process feelings."

I had never thought of the wilderness in quite this way: as a space for full physical presence, but it made sense.

And I could see it happening to Rochelle and now another participant, Kim, the youngest, at 20. "Being here and just feeling the breeze hit my face," Kim told me on the second night, "like feeling my nose go cold and feeling my fingertips go cold, like yes, it is really cold, and it's hard, but I can feel it and I can remember every single detail."

They were feeling, in a word, alive.

NIGHT FELL, just as cold. We made a crackling fire. There were songs, laughter, and more praying. Kris put a hand on my shoulder and asked if she could pray for me. "You shared with me a little about your divorce, and I've been there," she said. I was moved and a little embarrassed. She leaned her head toward mine. "I hope, God, that you will fill all the deepest, darkest crevices, all the broken crevices in her heart . . . and when she becomes confused or heartbroken and vulnerable, you will meet her where she is, overwhelmed with pain or sorrow. She will not feel alone . . . I thank you for what you have done for her and that we will be forever friends."

For a moment, the night felt less cold and less stark. The fire pulled us in; the group circled tighter. We repeated the drill of filling warm water bottles and bringing them to our sleeping bags. It felt like there'd been a shift in the group. We knew we'd be walking out in the morning on a beautiful trail. Because of the intense conditions, we'd be leaving the mountains a day early and spending our last night at a heated, well-stocked facility. This would help the women feel safe, and heard. I could sense the relief the Arctic-blasted trip was almost over, but also the joy of being together, of encouraging each other to continue the hard work of recovery. It wasn't just the glorious mountainscape that nurtured this, but their earned sense of being there for each other.

In the morning, we walked out as the wind kicked up, snow on its heels. Rochelle was once again feeling reflective. She took deep breaths. "Being here just sent messages to me like, wow, I actually do have to take care of my body. And I realized that I've never really claimed my body as mine."

I felt my own renewed commitment to building trust with others and to spending more time outside, partly for adventure and distraction, but mostly to find the beauty, peace, and space for healing now that I saw it working for these women who needed it so much.

As I'd walked through aspen stands and stared into campfires, I'd started thinking big. I was thinking of rivers. I wanted to embark on something whose scale matched the enormity of the hole in my heart. Per my conversation with Paula Williams, I needed to journey into awe. I needed to build some bravery. And I wanted something ambitious to plan and work toward to pull me into the future. I didn't have God and I didn't have a residential community of women undergoing the same journey, but I had a deep connection to rivers.

Passing the moose meadows on the way down, I fell into a conversation with Chelsea Van Essen about her work with trauma, and about our own heartbreaks.

"Heartbreak is trauma," said Chelsea. "Absolutely. It's trauma because it affects you on so many levels. It dismantles your identity." I was inclined to resist this idea of heartbreak as trauma, but the more people said it, people like Chelsea and Stacy Bare who worked with survivors of trauma, the more I began to accept it. I could see the way it was affecting my body.

Chelsea encouraged me to spend more time in nature and suggested I look into a form of therapy called EMDR, for eye movement desensitization and reprocessing. She told me she was co-leading a workshop in Denver in a couple of months if I wanted to check it out. I'd heard a little about EMDR, and I thought it sounded weird. But I liked Chelsea, and this workshop was designed specifically for people going through divorce. My people! I told Chelsea I would think about it.

"You know, eventually the brain will process these things," she said. "You're a high-functioning person, but it could take a year, or it could take a very long time." EMDR, she said, might well speed it up. That, I wanted.

11

SHAGGY BIRDS

Kiss my ass, I bought a boat, I'm going out to sea.
—LYLE LOVETT, "IF I HAD A BOAT"

Something had shifted for me during and after the Colorado back-packing trip.

Maybe it was the contagious resilience, or just the inspiration of being with brave women who had so much to teach me, or being in the mountain light and weather for a few days and recognizing my troubles were a small hiccup in the universe. But I was feeling something I hadn't felt for a long time: a sense of optimism. The world and its people were capable of displaying so much beauty and strength. I was still riding frequent lows, but I had to admit I was also riding highs (natural ones, mostly), and this felt surprisingly new to me. I had always just put my head down and gotten a lot done. I'd felt good about that, about my competent parenting, my lack of drama, my steady capabilities. But at what cost? Now that I was surfing a bigger sine wave of emotions, I unexpectedly . . . liked it? Even though the sadness was intense and uncomfortable, it somehow came attached to other feelings, of vitality and gratitude and even love for all the things that weren't out to hurt me. I liked having emotional and sensory range in my life. I liked the ability to feel things and I wanted more of that. It seemed both possible and necessary.

I still couldn't figure out how to settle down, not into a new notion of myself, and not into the idea of being alone. I didn't know who I was on my own, because I'd never met that person before.

"Remind me why it's good for me to be alone?" I asked my therapist, for maybe the third time.

"Being alone is like a muscle," said Julia. "One should exercise it, because you never know when you'll need it and you want it to be working."

Right. BUT. Plenty of people marry young and stay married and don't really ever exercise this muscle. Must I? Apparently yes, because I wasn't one of those people anymore. I needed practice.

I was scared of the future. Perhaps I was suffering from a failure of imagination, and what I needed was a better narrative. Helen Fisher had emphasized to me the power of story. To move on, she'd said, you needed to know, or think you knew, what had happened. In Utah, Paula Williams had said something similar with regard to the densely connected, meaning-making brains of people who are prone to experiencing beauty and awe.

In 2017, David Sbarra, the psychologist from the University of Arizona who had told me that divorce is practically synonymous with inflammation, published a study with colleagues titled "Tell Me a Story: The Creation of Narrative as a Mechanism of Psychological Recovery Following Marital Separation." They asked 109 adults who'd been separated within the past five months to complete a series of questionnaires and exercises, including expressive writing about their breakups. Researchers found that people given to ruminating often about their experience (guilty!) reported more psychological distress seven months later, and these individuals also had a harder time creating distance from their pain and being able to write about it coherently. When they were given specific writing exercises to make meaning from the experience, however—as opposed to just freewriting about their emotions—they later reported better psychological outcomes. The writing prompts included instructions such as Tell the story of the end of your relationship, Narrate the separation experience, and Describe an end to your divorce story.

The benefit of a story, Sbarra explained to me, is that it helps define who you are and it gives you perspective. "We start feeling better after breakups as we start rediscovering our sense of self," said Sbarra. "This is a core engine of recovery. A separation experience violates your meaning systems and the expectations you have for life, so how do you get things sorted out? How do you narrate the experience?" I told him that six months out I was still having trouble making sense of what had happened and what was going on. I mentioned I was forming plans for a big wilderness river trip, and he approved. Then again, he lives out west, and he understands the compulsion to seek solace in wild places.

"Being in nature is about expansion, and getting outside of yourself," he said. "The *doing* is the key. The ones who shut down and withdraw, these are the people with difficulty navigating the end of marriage."

The phrase *walking it off* explains a real phenomenon, according to neuroscientist Shane O'Mara at the University of Dublin and the author of *In Praise of Walking*. He explains that moving around can help prevent depression, as well as a host of arterial and metabolic woes. As blood pumps and new neuronal growth factors flow, we become more creative, more self-aware, more ourselves. "A simple, collateral effect of rising and moving," he writes, "is that activity spreads across more distant brain regions—increasing the likelihood that half-thoughts and quarter-ideas, sitting below consciousness, can come together in new combinations." I'd been walking a lot—four or five miles a day—but figured the idea should also apply to paddling, another long-distance, rhythmic, bilateral motion.

This was not a conscious goal while planning my trip—I just craved the activity. I had to move.

As plans for a long wilderness trip formed in my head, I reached out to Dave Strayer. A cognitive neuroscientist and professor at the University of Utah, he is a powerhouse in his field, but he's most content when camping out in the desert. A proponent of the so-called three-day effect I'd written about before—the idea that three days in nature can jumpstart our creativity and soothe our city-addled souls—he was someone I knew would be supportive. As a river runner, he would also, I knew, offer some practical advice.

When he visited Washington, DC, for work, we met at a celebrated restaurant. As happily as Strayer can exist on Pop-Tarts and ramen in a campground, he and his wife, Kay, are world-class travelers, gourmands, and oenophiles. Like me, he straddles a pretty extreme nature-city divide.

Sitting on a patio surrounded by hanging plants, I told him I wanted to find a river long enough to sustain many weeks of paddling, some with friends and some all by myself.

He raised his eyes.

"A solo?"

"Yes." I'd been thinking hard about this, about the need to find courage, to learn self-reliance, and to confront my dread of being alone. After 25 years of marriage and many springs and summers paddling with either my father or my husband, I needed to learn, both literally and metaphorically, how to paddle my own boat.

"Interesting," mused Dave, pouring us more Riesling.

We discussed the pros and cons of rivers we both knew and had run before in Utah, Colorado, Montana, and Idaho. A logical candidate was the Salmon, because it is free flowing, wild, and crazy beautiful. Once scouted and then abandoned by Lewis and Clark, the Salmon runs through country that has burned up in big wildfires over recent decades. Yet now its banks were coming back, vibrant and green, an irresistible metaphor. But the Salmon's rapids are tricky, the current fast, the climate often cold and wet. The desert rivers, on the other hand, can be too hot and, while gorgeous in places, are plagued by stretches that are windy, buggy, and flat.

"Do you want to look inward or do you want to look outward?" he asked, almost rhetorically.

"Um, inward?" But I also wanted adventure. Per the lessons of Chelsea Van Essen and the other trauma experts, I wanted comfortable adventure, if that wasn't too much of an oxymoron.

"Oh, if you're alone, you'll have adventure," he said. He confessed he'd spent only one night alone in the wilderness. "That was enough," he laughed. And he was a Boy Scout leader. I had never been out alone.

"I want you to come back in one piece," he said. "I think the desert would be better." Then he quoted the writer Terry Tempest Williams. "Every pilgrimage to the desert is a pilgrimage to the self."

The Green is the largest tributary of the Colorado River. It starts as a trickle in the snowmelt of western Wyoming, but gathers plenty of force as it carves its way through canyons in northwestern Colorado and then makes a long, lonely descent south through most of Utah before merging with the larger river a ways above the Grand Canyon. I'd run parts of it before, the broiling, fun Lodore Canyon, which takes about four days to paddle, and Desolation Canyon, a weeklong trip I first canoed with my father. But I'd never ventured south of that, and Dave had. To start in Wyoming and paddle most of the river would take at least a month, he said. It would make the most sense to run the more challenging whitewater stretches with other people, both for reasons of safety and to help me carry supplies so I could paddle a light, small boat. But for the last two weeks, the river slows considerably as it flows through two spectacular canyons above the confluence with the Colorado River. I could switch to a solo canoe big enough to carry all my own water, food, and gear. There's only one road that reaches the river in that stretch, and it would be best to have someone meet me there for a resupply halfway through.

"I could probably help you with that," said Dave. The road to Mineral Bottom is dirt, unpredictable, and a good five hours from Salt Lake City. We looked at our calendars. July and August were the months that worked best for me and the kids, since I wanted them to join me for sections. Assuming I could even secure the necessary camping and river permits, he would likely be on sabbatical halfway around the world then. He considered. "My grad students like you. Maybe they'd do it?"

I SPENT MANY hours over the winter and spring talking to river people, poring over maps, and cajoling friends and family members both with and without their own boats to join me at various points. I started thinking about food, equipment, the risks of going into a roadless wilderness without any insulin as a budding diabetic. My doctor thought I could put off the need for medication for a few more months if I ate carefully. With a good cooler, I could pack in fresh meats and vegetables and avoid high-carb camping food.

And then there was the solo piece. How would that be, alone in the desert, for weeks? I had no clue. I called one of the adventure therapists I'd met on the trafficking-survivors story, rock climber Aleya Littleton. She said, wisely, that undertaking a solo is not the same thing as being alone, because—ideally—you feel accompanied by the love and support of friends and family, as well as by the lively presence of the natural world. I would have a loose ground team; I'd make friends, in a sense, with some new creatures—mammals and birds—and I would have the time to enjoy them.

And yet, the solo piece would be significant. "All your mistakes and your accomplishments out there are your own," she said. "You can't blame or credit anyone else. It's a reminder or a reacquaintance of what it is to be you."

I THOUGHT ABOUT Emerson's writings on nature, some of which were (and are) gently mocked. He rhapsodized about the transfusion of mind and nature, even "the suggestion of an occult relation between man and the vegetable." But the most influential American philosopher of the nineteenth century also said this: "In the woods, we return to reason and faith. There I feel that nothing can befall me in life,—no disgrace, no calamity, (leaving me my eyes,) which nature cannot repair." I revisited his famous essay "Self Reliance." Written in 1841 when he was 38 and living with a houseful of children and obligations, it argues that community is a distraction from self-growth. Social visits, family demands, expectations of conformity, formal religious institutions—Emerson wanted little of it. Under such influence, he writes, "we come to wear one cut of face and figure, and acquire by degrees the gentlest asinine expression."

Emerson helped give language to the American strain of self-definition and rebellion. Many men and women would take up the cry, rooted in European Romanticism, to seek secular time in nature to cast off the strictures of society. Wordsworth and Rousseau did it, John Muir, Mark Twain and his Huck Finn did it, and so did Simone de Beauvoir on her treks through the Alps, Jack Kerouac ("Climb that goddamn mountain"), Wallace Stegner, William Least Heat-Moon, Chris McCandless, Cheryl Strayed, and scores of other misfits, naïfs, adventurers, artists, and repiners.

"It should not be denied . . . that being footloose has always exhilarated us," wrote Stegner, channeling white men especially, " . . . and the road has always led West."

Now that I was single in middle age, and therefore officially nonconformist, I knew I should try to reframe my circumstances as having certain advantages. I was, after all, learning more about my edges and desires, sexual and otherwise. I felt, at times, uncaged. It felt dangerous, not always in a terrible way. I wouldn't have expected any of it. Of course, people who exist outside of norms are drawn to ideas like self-reliance. The idea that hard times can lead to intense personal growth isn't new, but I found myself particularly curious about the studies on creative insight.

In 2013, an organizational behaviorist at the Johns Hopkins Carey Business School named Sharon Kim and two colleagues wrote a paper called "Outside Advantage: Can Social Rejection Fuel Creative Thought?" In it, they argue that the negative consequences of social rejection are not inevitable, but rather "depend on the degree of independence in one's self-concept." Just as the need to belong is a strong urge, so is the need to individuate, and for some of us, that need is particularly strong. In these cases, they wrote, social rejection might amplify feelings of distinctiveness as well as creativity by "conferring the willingness to recruit ideas from unusual places and move beyond existing knowledge structures."

To test the idea, the researchers asked 43 volunteers to take a psychological questionnaire assessing their "need for uniqueness." Then all the volunteers were privately told they had been rejected from a group task, so they needed to complete it on their own. It was the RAT, which stands for Remote-Associates Test, and it's commonly used by psychologists to measure creative thinking. In it, each respondent must solve a bunch of word problems in a limited period of time, thinking up one word that connects to three seemingly unrelated words. Example: *fish*, *mine*, and *rush*. (Answer: *gold*.) According to the paper, "less creative individuals perform worse because they are biased toward high-frequency (common, but incorrect) responses."

In the experiment, people who evinced a higher need-for-uniqueness on their questionnaires performed significantly better on the word problems than others. And yet, this would seemingly contradict other studies showing we lose some cognitive skills after being rejected. Maybe, as Paula Williams suggested, individual differences in personality really matter.

"This is not to suggest that rejection is necessarily a positive experience," said the authors. The high-unique volunteers still didn't like being rejected. It still felt bad. But for whatever reasons, it may have motivated them to get their heads off the pillow and prove how individualistic they were.

When I called Kim to ask some follow-up questions, she began with words of caution. "I would never suggest one's self-concept is immune to the effects of loneliness and rejection," she said, "but what our study suggests is that there are ways to maybe channel that experience, to maybe generate ways to make the best of it, especially if you are one of those people who already think of yourself as being different."

Did I? I wasn't sure. Heartbreak, of course, is a much more complicated experience than being booted from an afternoon group exercise. Still, elements of her team's analysis made some sense to me. "At the heart of what we think is happening is that you see yourself as having independence, and there's a quality of finding strength in that individualism," she said. "To the extent that you can leverage that under these circumstances, and see some positive consequences in describing yourself in new ways, that could be helpful, for example, not essentializing yourself as married or divorced. You can change the rules for what it means to be accepted or rejected and make different rules of belonging."

Married. Divorced. Dumped Girl. I was tired of the plangency of it. It was time for a new story, if I could create it.

OUT WEST, I visited a dear friend, whom I've known since we first attended summer day camp together at age seven. Mara's a physician who grew up with a traditional education on the East Coast, but she's learning that Western medicine isn't always enough to make her patients feel well.

One night over dinner at a restaurant she gave me some hard, clear-eyed love. She said I'd failed to hold my husband and myself accountable, that I was complicit in the marital death march.

She congratulated me on going for it with Ennis and the Apocalyptic Poet. That was more the old friend she knew, the one who ventured forth and looked for what she wanted.

She was right about my marriage. Its flatness had seemed a reasonable trade-off for security and family. But if Utah's Bert Uchino was right, keeping the peace requires too many withheld emotions and sublimated needs. Mara was reminding me that I had dimension beyond that flatline.

She told me I'd been a great friend to her, that she loved me, and that while my life might currently look like a toxic Superfund site, she was glad for the opportunity to help me through it. That's when I started crying, hard. The waiter avoided us.

She gently suggested I try meditating.

"I do try! It's hard!" I wasn't very good at it, especially with my head on high alert. She said she would talk me through what she called a "wise guide" meditation. So when we got back to her place, I lay on the shag rug in her basement. I was skeptical, but I was feeling relaxed from the wine or from the good cry. Shag is so underrated. We breathed deeply for a while.

"Imagine a safe place," she said, and was quiet for a long time. I pictured the shoulder of a mountain, green and solid, the sun shining. Small white clouds floated above me. My kids were nearby. "Now imagine a benign presence, something or someone arriving to offer you wisdom."

I had a clear vision of a great blue heron flying in close.

"This presence is telling you something."

This was the bird I often saw on rivers when I was a child and more recently in my regular tromping around the Potomac River in DC. This bird enjoys standing fetchingly on the shore, still and impervious, like a cat who pretends it has better things to do. When you get close it alights in an improbable jumble of limbs and wings to another perch just downstream.

I opened my eyes and told Mara what I'd seen.

She looked up the symbolism of the heron. "If Heron has come wading across your path, it is time to look deeper into aspects of your life that will show you how to become self-reliant," she read. "A heron signifies your ability to explore."

Emerson himself couldn't have conjured a better animal. Self-reliance. Exploration. Keep moving. It was perfect.

12

THE WIZARDS OF LONESOME

I believe you are standing in the place
where I am supposed to be standing.
—LEONARD COHEN, *BOOK OF LONGING*

I didn't want to take just acupuncturist Bernie's word for it, so I accepted
Steve Cole's invitation to fly out to Los Angeles for a molecular analysis of my blood. We might be able to see what shock looks like.

At the UCLA Social Genomics Core Laboratory, a phlebotomist
siphoned off four milliliters of my blood into a special vacuum-sealed test
tube designed to keep it stable in deep freeze until it could be analyzed
with a larger batch of blood from other study subjects. Bandaged and
coagulating, I walked a couple of blocks to the Louis Factor Health Sciences Building to meet Cole in person. Not many structures on campus
rise above a few stories, but Cole is lucky to perch on the 12th floor, where
his small office overlooks Bel Air.

Cole may be doing pioneering, even radical, work, but his appearance veers toward golfer. His face is smooth and ruddy, his hair very short, his speech midwestern. On the day of my blood draw, he was wearing khaki pants and a blue checked shirt.

We talked about my idea for a summer wilderness expedition, and what blood draws at different time points might reveal if we captured another sample shortly after the trip. It would be interesting to see what, if anything, had changed. I was, by this point, pretty invested in the idea that nature could help me heal. Cole was game. "Your vision of the experiment sounds right to me," he said, although he had never looked at nature immersion as a way to treat loneliness.

He had examined people's blood after other interventions, like daily meditation. As Cole explained, you can take all the questionnaires you want and burble on about how much better you're feeling after, say, a yoga class or a walk in the park or getting engaged. But the real question is whether your body—your immune system, to be precise—is still gearing up to face a threat that doesn't really exist. If so, you're not doing as well as you think. He and Bernie had essentially arrived at the same insight through very different paths: the body knows.

The cells that enthrall Cole are monocytes, which make up about 5 percent of our white blood cells. Monocytes are the ones that closely monitor our nervous system. They are attentive to our moods. These cells are deeply evolved, also found in fish and many other lower-order organisms, adjusting the immune system based on threat and uncertainty. You want to produce lots of wound-healing inflammation if you are headed into the maws of a giant squid.

Cole's blood analysis would extract RNA, the messengers that take information from our DNA to make proteins that carry out biological actions in our cells. Because RNA is constantly transcribing messages, it provides a real-time snapshot of the machinery renewing our bodies. The white blood cells of our immune system die after several days or weeks and need to be remade in our bone marrow. "When you're stressed," said

Cole, "the bone marrow gets the instructions that say you are about to be injured or attacked, so it shifts priorities." It might make more monocytes and fewer B cells, for example, providing an individual snapshot of your threat radar in the days before your blood draw.

Here's the chronology: We sense a threat. Our brain's fear center, the amygdala, instructs our sympathetic nervous system to release the fight-or-flight neurotransmitter norepinephrine into just about every organ of our body. Our monocytes carry receptors that allow them to sense the norepinephrine and fire up the production of RNAs that make inflammatory proteins, while also shutting down the production of RNAs that defend us against viruses. At the same time, sympathetic nerve fibers in our bone marrow prime our stem cells to make more monocytes (and inflammation) and fewer lymphocytes for fighting viruses. As Cole later would put it, "It's a recipe for Covid, not to mention heart attacks and cancer and many other life-shortening maladies of loneliness."

There are lots of ways to experience shock and misery, from loneliness to PTSD to discrimination to hatred of your job to physical pain, and they register in similar ways in the genome, Cole explained. He zeroes in on a block of about 200 genes that make up what he calls "an organized conspiracy" in the immune system, joining forces to launch an uptick in inflammation and a down-tick in our ability to fight viruses. Up goes the production of molecules like C-reactive protein, interleukin-6; down goes the production of antibodies and protective proteins like type I interferon.

The immune system can't protect us against everything all the time. It has to make strategic decisions. Cole's theory is that it makes a snap judgment to assume that when we are lonely (or threatened in any other way), we are more likely to need inflammatory protection against biological assaults and bacterial infections. The body's reasoning goes like this, he said: alone and without the protection of others, we are more likely to acquire a serious physical wound, and less likely to catch viruses, which are spread in groups.

In his HIV studies, Cole found that type I interferon molecules, which drive the antiviral response, were deeply suppressed in people lacking social support. Unfortunately, this was exactly the wrong call for fighting HIV.

Our bodies, as Cole puts it, are quite adept at turning misery into death. When we are stuck in a state of fear, inflammation becomes a powerful fertilizer for cardiac diseases, neurodegenerative diseases like Alzheimer's, and metastatic cancer. Cole goes so far as to call our RNA's response to threat "a molecular soup of death. It is a big recipe for how not to live."

As for my own experiment, Cole suggested that we take more than two time stamps of blood. The more the better, he said, for gathering a clearer picture, and he was curious to see if more time elapsing since the breakup would yield trends in gene expression, as one might hope.

As usual, Cole ended our meeting with a pep talk. People who live alone are not doomed, he told me. Plenty of single people build deep friendships, fulfilling lives, and strong support in case they get into trouble. Their immune systems may not feel under threat at all, while married people with lots of friends may feel like no one is looking out for them. I had meaning and purpose in my life, he pointed out, or else why would I be here trying to make sense of a smudge of monocytes to share with other people? I had kids and friends who loved me. I had a plan for a big river trip.

And for those times I forget all that? "Try getting mentally zen," he said.

LONELINESS MIGHT REFLECT our emotions, but it begins in experience. Modern life has certainly made the experience more common. As Cole's late colleague John Cacioppo wrote, "Western societies have demoted human gregariousness from a necessity to an incidental." For decades now, we haven't been passing time together as we once did, not in romantic dyads or arranged dyads, not in kin groups, and not even in groups of friends or neighbors or in civic groups. More than ever around the industrialized world, people are living alone, increasing the risk of *feeling* isolated. In the US, single-person households have quintupled over the past five decades to 36 percent. In 1900, 2 percent of Americans aged 45 lived alone. Today, it's nearly 10 percent. Worse, a third of all Americans over 65 live alone.

People are living longer after they lose a partner to death or divorce. They are aging with more illness, too, and so are less able to get around and visit friends and family. But it would be a mistake to relegate loneliness to the elderly. People aged 18–24 are more likely to say they feel lonely than any other group, according to recent large surveys in the UK and the US. They are marrying later, having fewer children, sometimes not marrying at all. Much of this is by choice. But across industrialized societies, millennials are forming fewer long-term relationships and having less sex overall. Loneliness rates in much of the Western world stand at about 25 percent. Several months into the coronavirus pandemic, rates had risen 24 percent above that.

There's no doubt that social isolation and loneliness feed off each other. Confoundingly, people who are deeply lonely can be changed by the experience in a way that can make it harder to emerge from it, as with Zelikowsky's mice. They become more suspicious of others, less socially confident, more easily wounded. It becomes easier to retreat.

Loneliness matters not just to individual health outcomes but to political systems. Consider the observations of psychologists Richard S. Schwartz and Jacqueline Olds, who argue that social isolation makes people less empathetic and drives them to view others not like them as threatening. They may be more likely to call for closing borders, for example. Loneliness further contributes to political polarization by driving us to seek affiliation with ideological tribes in lieu of social ones. In fact, the rise of loneliness in both the US and the UK corresponds to a surge in extreme politics and nationalism, leading to the demonization of outsiders, as well as to the rejection of policies that promote fair trade and human rights. And the lonely Left may be just as tribal as the lonely Right. It is a mutual cycle of retrenchment and distrust.

Around the world, especially in wealthy countries, people are dating less. Japanese adults have the least sex overall. Although pornography is popular there (and sex dolls increasingly so), the Japanese have a phrase, "herbivore men"—*soshoku danshi*—for those who are not interested in flesh, or rarely date. They also coined a word for elderly people who die alone in their apartments (*kodokushi*) and another for young people who refuse to leave their rooms (*hikikomori*). This is not to single out Japan as highly unusual; based on global trends, the country represents the future.

In the UK, 41 percent of Britons say the TV or a pet is their top source of company. The country ranks second in the world for percentage of single-person households (Sweden is first, Japan is fourth, and the US comes in fifth). About 20 percent of British adults report feeling lonely most of the time, and they're more likely to feel this way if they live in a city. Officials there are alarmed enough by the public health consequences that in 2018 they designated a new parliamentary post: minister of loneliness.

Perhaps it's fitting that I decided to pay the minister a visit during Thanksgiving week. I was bereft not to have custody of my children during the holiday, so when a landscape institute invited me to come to England to give a talk, I said, *Yes, please*.

Before making my way to Parliament, I stopped by the office of Andrew Steptoe. A psychobiologist and epidemiologist at University College London, he is the director of the English Longitudinal Study of Ageing, a large and revered survey of social factors and health, which has found, among other things, that loneliness and social isolation predict a decline in verbal fluency and other cognitive skills, and increase the risk for heart disease, stroke, and depression through dysregulated inflammation, cortisol, and sleep.

Steptoe waved me to a comfortable blue chair in his office off

Tottenham Court Road. Primed for sociability, the room was filled with seating options and a round table for team meetings. Pictures of his grandchildren lined the walls. Now in his late 60s, Steptoe wears a Warholian curtain of white hair across his forehead. "We've been banging on about social isolation for a few years now," said Steptoe, who is pleased the government is taking it on.

He was cajoled into studying social factors and health by his late wife, the influential psychologist Jane Wardle. After studying the effects of social connections on cancer prevention and outcomes, Wardle herself succumbed to leukemia in 2015. I noticed Steptoe still wore his wedding ring, because I notice these things now. I asked him about it. Even though he knows well—all too well—the importance of trying not to feel lonely, he said he doesn't always succeed.

"It's difficult to get over," said Steptoe. "No matter how many social contacts I have, when you get back to your flat, there you are."

But Wardle left her husband the legacy of her research, including an understanding of the importance of purposeful work and hobbies. It's a message he intends to spread. "I value those aspects more than I used to," he said. "I like to help people, and that's become truer."

THE HALLS OF Parliament did not feel particularly collegial. The government was tumbling through its Brexit-induced freefall, but it happened to be (what were the odds?) Loneliness Awareness Week. The hallways were littered with pamphlets, buttons ("Let's Talk Loneliness"), and posters. I was greeted in the lobby and escorted upstairs to meet Conservative MP Mims Davies, minister of loneliness. With a title like that, I half expected her to wear a pillbox hat and carry a magic wand. When I entered the spacious and sun-drenched office of the minister herself, she apologized for her tight schedule and gestured me to a sleek couch beneath a large window.

She told me she'd been appearing at a series of luncheons all week to announce a new £11.5 million initiative designed to showcase the problem and inspire more people to join community programs. "Not having your social connections can be as bad for you as obesity or smoking," said Davies. "I think it's quite easy to think it's just a rural isolated older person's issue, but it's not. You might be a young person in a new city. You might be a caregiver. You might be a new mom. You might be someone looking after somebody with a disablement. But also if you can recognize it in yourself, or from your previous points in your life, you can recognize it in all the people around you. You can help and reach out to people."

The minister of loneliness wore a simple white jacket, black shirt, and gray-and-black animal-print scarf hanging loose. Her hair also hung straight, blond and wispy, shoulder length.

"Do you personally feel that you can relate to this issue?" I asked.

"Absolutely!" she said. "I was one of the first people in my group to have a baby and everyone else was out having a great time and I was stuck at home." She told me her kids were now 9 and 14. When she first got to Parliament several years ago, she didn't know anyone and people weren't very helpful. She would get lost in the hallways. Her dad was ill for a long time, and her mother had to care for him without a lot of support. Soon she was ticking off more life events: both her parents died, and her own marriage fell apart.

About the divorce, she said, "Everything that's around you that you're connected to is gone as well. So you've got to restart. There's no road map for you. And now both of your parents are gone and, you know, you start to recognize who's there for you." Speaking up was her mission: loneliness must be revealed. "Nobody walks around with a big arrow on their head saying, 'I'm lonely.' We want to take the stigma away. It's very human. It's a human emotion like anything else and yet so hard to talk about."

To help people become more socially engaged, the UK has some very UK-sounding ideas, like turning pubs over to knitting groups during the morning hours, and creating "chatty tables" at community centers, and even "chatty buses," which I didn't get a chance to ask about before time was up. There are rambling groups, bird-watching groups, amateur choirs, book clubs, community gardens, baking groups, and something called a "befriending service." Some of these options are even written up for lonely people as "social prescriptions" by medical professionals.

The MP's assistant had already lined me up to visit a program the country was quite proud of. Which is how, on my first free day in England, I boarded a train for Redditch, a suburb of Birmingham.

ROLAND DUKE GREETED me outside the Redditch station. He approached with an apprehensive stiffness, as though walking through a wind tunnel. We folded into his small car for the short ride to a low brick building. "I'm not sure where I heard about men's sheds," said Duke, who is 70. "Something on the telly, probably." A stocky retired firefighter, he told me he didn't want to just sit around at home, and his wife didn't want him to either.

He created this "men's shed" in a former industrial garage a couple of years ago, with some funding from a charity called Age UK. Later, he got additional help from the national lottery. Part of a movement that originally started in Australia but has since spread to many countries, the sheds aim to bring men together to share power tools, work on small projects, and, at least in Britain, drink massive quantities of tea. But the stealth idea is to combat loneliness. "Women are more social, they'll go out and have a chat, and blokes don't," said Duke. "That's where the shed comes in." Men, it is sometimes said, prefer interacting with each other at indirect angles—shoulder-to-shoulder while doing something—rather than face-to-face. "If there's a comfortable way for them to talk," he said, "that's good."

In the shed, there's no pressure on the men to converse; there's no group therapy leader or social services questionnaires. But they know what the shed is for, and they like the place. It's open for three hours in the middle of the day, three days a week. "We have widowers, men with cancer, a young bloke, 40, whose mum died," he said. There's the guy who almost lost his hand and the one who was in a motorcycle accident and the man whose girlfriend died in a car wreck 30 years ago who never quite got over it.

He opened the door into a drafty, chaotic space. It was like walking into a Western saloon. The chatter stopped and about 10 (mostly gray) heads turned to me. They nodded, and then the talking and the tools started back up. Duke took me over to meet a gray-bearded man in a blue fleece jacket. Geoff Franklin was lining up two corner pieces of wood for a display box. "I'm really only here to keep out of mischief," said Franklin in a West Midland Brummie accent (Ozzie Osbourne's dialect, if that helps). "I was an engineer all me life. I like making things." In his mid-60s, he's a widower with no children. "I used to make nuclear submarines," he told me.

"Did ya have security clearance?" asked Duke.

"I can't tell you!" said Franklin, to much laughter.

"This is the company of like-minded people," Franklin said to me. "They can see what you're working on and say, well, you really made a mess of that! But there's no rancor. We all come for different reasons. I rolled up one day and Roly said, 'Well, why don't you have a cup of tea?' We just all hang out." His words got drowned out by the roar of a planer, so Duke waved me to another corner of the long room, lined with sawdust-covered work benches, machines, and pieces of lumber.

There, Roger Harris was studiously cutting a folded piece of paper. Duke introduced him like a proud papa, telling me he had recently made a viola here for his grandson out of maple and spruce wood. "I'm over the moon he can play it. He's 14," said Harris, a large man with Captain Kangaroo white hair and mustache, wearing a red wool shirt. Most of the projects are not so ambitious: bat boxes, shelves, and bird boxes for a local schoolyard. Harris explained that today he was making a paper snowflake that he learned about on YouTube. He was also spending time helping someone use the pillar drill. He comes faithfully twice a week, whether he's working on a project or not. "We always count the fingers before and after the day," he bellowed.

I met another man, whose wife deposits him here on her way to aerobics at Slimming World. This was like adult kindergarten, complete with snacks and drop-off times. Some men make the decision to come on their own. A tall man named Arthur Dyas told me he wasn't working on anything today. He was just here to chat. He comes often. "There wasn't another way I would have made new friends," said Dyas, a pensioner who lives alone. "It's not on the agenda. I do have old friends. I also go to the same pub every Friday night, but this is cheaper and you can drive home afterwards."

Dyas said he just returned from a conference all about men's sheds, which now number 472 in the UK. He leaned toward me conspiratorially, saying, as if I didn't already know, the woodwork was just a smoke screen. "The real stat is over a million cups of tea drunk." He passed me a brochure with more data: 52 percent of shedders gain a renewed sense of purpose, 24 percent feel less lonely, 75 percent experience a reduction in anxiety, 89 percent a decrease in depression, and 97 percent make more friends.

"I make friends easily," a 78-year-old named George told me.

"Who told you that?" bellowed Duke, and everyone laughed.

I could see why the men enjoyed it here, with the shed's whir of machinery and the genial vibe. I couldn't help noticing, though, that all the shedders were white as King Arthur. The like-mindedness, as one of them put it, is part of the appeal. But what about people who are even more marginalized, not just by age and social circumstance but by race and sexual orientation and physical or mental capabilities? As I would soon learn, these are questions the government is also starting to ask. Sometimes finding community means finding it with people who look like you, but it shouldn't, even in rural England.

I guess I looked out of place, too, or maybe these men have a keen eye for spotting loneliness in others, such as American moms traveling alone during Thanksgiving. On my way out, Roger Harris came over and held out his paper snowflake. I thought he was displaying his handiwork.

"If I give this to my wife, she'll have me keep making them for the Christmas tree," he said. "You take it."

PART THREE

AWE

13

TRUTH SERUM, PART ONE

It is stunning, it is a moment like no other,
When one's lover comes in and says
I do not love you anymore.
—ANNE CARSON, "THE GLASS ESSAY"

Into late fall, Ennis and I were sharing long phone calls. In one, he told me he deliberately tries to date newly separated women. *Because they're so easy.*

"No!" I responded. "Stop doing that! We are like skinless puppies."

"Really?" he asked.

It was hurricane season. Puerto Rico lay underwater and a gunman shot 471 people from his room at a Las Vegas hotel. Tom Petty died and Ennis's favorite songs—the ones he sang to me—played in an endless loop on the radio.

I got back new lab results of a three-month blood-sugar measure. I was now a tenth of a point away from the official diabetes red zone. I often felt nauseous, and lonely. I still wanted someone to hold me and send notes like Virginia Woolf sent to Vita Sackville-West in 1927: "I love your legs and I long to see you."

Despite the tensions of divorce mediation, my husband and I were getting on fairly well. The handoff of children and dog was going smoothly. The kids were so game, such troopers, packing up their stuff every week, juggling two sets of keys. I found ways to spend extra time with them on their father's weeks, like picking up my daughter from school and bringing her back to my house to do homework until he got home after work. I brought the dog back to my home office to hang out with me while everyone was gone for the day (the separation was hard on her, too, I figured, but I also liked having her big eyes following me around all day). Sometimes the kids and I walked her together after school. If my husband had a work dinner and I was in town, he let me know and I'd take the kids out or bring them to my house. He, in turn, picked up slack if I needed an extra day on the road. He was an attentive, involved parent. Seven months out, we'd found a rhythm.

I was wondering if I even needed Chelsea Van Essen's divorce EMDR workshop in Denver, but then I figured maybe I could put some of it to work defusing my still-voltaic feelings around Ennis. I arrived at the starkly lit basement of a 1980s-era community center, a place that had no doubt seen a lot of AA meetings. I was a few minutes late, and the others—six women—were already seated at desks facing a whiteboard. On the desks lay pieces of blank white paper and neat rows of crayons. It looked like preschool for sad people.

We introduced ourselves and talked about why we were here. A trim and fit woman around 60 had been married for 34 years to her husband, a dentist. She'd recently found out he'd been sleeping with the hygienist for eight years. He wanted to stay together and she was like, *Uh, no.* Another participant had been married for 20 years, and although it was her husband who wanted out, their teenagers were blaming her for the split. Several others had left their husbands and were dealing with their guilt.

Kelly Smyth-Dent, a therapist and our workshop leader, welcomed us and introduced Chelsea, who would be assisting. In a dress, boots, and lipstick, she looked different from her trail persona but still had that great unhinged curly hair and wide-open smile. Smyth-Dent, with a stylish blond bob, appeared warm and professional. She told us we'd made a good choice to be here. EMDR, which stands for eye movement desensitization and reprocessing, is still a very new treatment for helping people dealing with traumatic memories and other adverse life experiences. Typically, it's applied by a therapist working with one patient at a time, but some clinicians have been experimenting with using it in group settings in communities that have been affected by traumatic events such as forced relocation or natural disasters.

EMDR is supposed to reduce distress associated with memories. "The point of EMDR," Smyth-Dent told us, "is that it releases heavy emotion without judging. EMDR is like cleaning out a wound." After treatment, a scar remains, but ideally it's no longer suppurating. The premise makes some sense, but in practice it looks pretty bizarre. In typical one-on-one sessions, the therapist sits directly in front of the client, moving a finger back and forth like a wide metronome across the client's field of vision, all while (or shortly after) the client is conjuring a particular, painful memory. This may last a couple of minutes until there is another memory. In group sessions, the therapist directs the participants to fold their arms over their chests, sarcophagus-style, and then to alternately tap their left and right hands rapidly and gently below their collarbone.

Why is this worth trying? Because when we're still constantly reliving the moments of trauma, our bodies stay stuck in the zone of threat-based hyperarousal. Lingering in this zone is, as we've established, bad for our health. It's also bad for basic functioning, because it can interfere with our ability to make considered decisions, regulate our moods, and feel empathy for others. We become insufferably self-absorbed because we

feel our personal survival is at stake. When we calm down, the real healing can happen: the emotional growth, cognitive insights, planning for
the future, and ability to connect with other people in reciprocal, meaningful ways.

Chelsea took over the next segment, talking, as she does, at a rapid
clip. She presented a PowerPoint filled with images of the human brain,
to show what happens to neurons when we are besieged by overactive
memories of stressful events. I found her schematic explanation a helpful
reminder of why it's so hard to just get over ourselves. The neural war zone
is governed by the amygdala, a small, Milk Dud–shaped cluster at the
oldest, lowest part of our brain stem. Chelsea described it as our brain's
smoke detector, but it's often called our fear center. When it receives a
threat signal, it unleashes instructions for our respiration and heart rate
to speed up, the blood to leave our digestive and sexual organs in favor of
our large muscles, and our vision to narrow, all in preparation for possible
escape or combat. All this may happen before our top brain—our "thinking" gray matter—is even aware of the threat, for the obvious reason that
in some dangerous situations there is no time for reasoning.

The deep-brain hardware is pretty crude. It doesn't easily distinguish
between a physical threat (snake!) and a social threat. It wasn't built for
the nuances of social interaction, love, and caregiving. The parts of our
brain that govern emotions are found in the mid-brain limbic structures.
And the cerebral real estate that manages judgment, decision-making,
logic, culture-creating, meaning-making, and communication lies in our
outer brain layers, the more recently evolved neocortex.

The challenge for us, the brokenhearted, is to try to keep the amygdala from stealing all the oxygen in the room. Anything that helps us
calm down will divert some juice to other parts of the brain so we can
rest, digest, reason, love, and grieve again. We need to disengage the fight-
or-flight response when we are not, in fact, in immediate need of fighting
or fleeing. Although it appears everywhere today, the term *fight or flight*
was coined by the psychologist Walter Cannon in 1915. Interestingly, he

applied it to stress derived from interpersonal relationships as well as from physical threat, noting that either way, adrenaline activates the sympathetic nervous system in preparation for a violent burst of energy.

EMDR promises to dial down the acute agitation brought on by specific memories. There seems to be something grounding and reorienting about the bilateral stimulation—tapping left, tapping right, tapping left, tapping right—that helps place the memory in context so that the patient can move past it and reimagine a better future. The memory becomes just a story, not a relived experience. I wanted my pain to dissipate the way Alain de Botton describes heartbreak in *Essays in Love*: "My love story was like a block of ice gradually melting as I carried it through the present."

As Chelsea had told me on the backpacking trip, the ideas with wilderness therapy and movement therapy are similar—to take the storm energy out of the fear system and disperse it through the sensory and motor systems and eventually through the cognitive network where bad memories can finally be made sense of in a more manageable way. Moreover, this dispersal can produce a happier vision of your future self. "Seeing novel connections is the cardinal feature of creativity," writes trauma expert Bessel van der Kolk. "It's also essential to healing."

THE FIRST PART of the EMDR treatment involves finding the right memories to defuse. This requires some excavation. After Chelsea's presentation, Smyth-Dent asked us to draw a time line of our breakups: the salient events and conversations that got us to our current state. Among other things, I identified the horrible morning when we told our children we were splitting up and the day my husband rolled his suitcase out of the house forever. Since my time line up to the present moment also included Ennis, I marked the spots on the time line when he bruised me more in ego than in body and the conversations when I learned about his harem and, later, about his conveyor belt of divorced women.

Next, we had to select one of those events and access our feelings around it. Here's where the fat crayons came in. Because we were in a group, we were asked to draw our feelings, not describe them. And with clunky crayons, "it's easier for people to draw really hard and fast if they feel angry or intense," Smyth-Dent explained later. I circled an inciting event in which I first learned about my husband's emotional attachment to someone else.

I drew a bunch of ragged red lines in a chaotic whorl against a background of black shadows. After a couple of minutes, Smyth-Dent asked us to scan our bodies. How did we feel? Pretty horrible. We were to write down a number between 1 and 10 that described our "subjective unit of distress," with 10 being the worst. I rated my distress at that moment as a 7. Then she told us to keep observing our body as we hooked our thumbs together and placed our palms across our chests to begin the butterfly tapping. This lasted a couple of minutes. It did feel somewhat grounding, or at least distracting, which may be the same thing. After that, we were to draw again how we felt. I drew a rough figure of a woman with red funnels around her throat and lower abdomen, where I felt physical heaviness and constriction. I ranked my distress as a 6.5, and we tapped again. We repeated the sequence twice more. The scratchings became looser, with more blues and blacks, and by the end of it, my distress number was 5.9. Then we were asked to draw a vision of ourselves in the future and tag it with one word. I sketched a woman on a mountaintop surrounded by blue. She was wearing a sun hat and her arms were starting to open out. My word was *Love*. Many of the women were crying as they drew.

SEVERAL WELL-REGARDED STUDIES have compared EMDR therapy to other forms of treatment. One found that three 90-minute sessions relieved symptoms of PTSD in 70 percent of EMDR-assigned patients, compared to 12 percent of patients assigned to prolonged exposure therapy. In another randomized study of 88 patients with PTSD, 30 percent of them became asymptomatic after eight weekly sessions of EMDR, compared to 15 percent receiving eight weeks of Fluoxetine (Prozac) and

12 percent receiving a placebo pill. The results were even more dramatic six months later, with 57 percent of the EMDR group reporting a total loss of PTSD symptoms compared to zero patients receiving medication. The study noted the EMDR treatments were not as effective for patients whose trauma experiences began in childhood.

This does not necessarily mean that the eye-movement therapy itself does the trick. While some critics acknowledge that EMDR can be effective, they argue that is because the therapy employs a similar technique to a version of cognitive behavioral therapy that repeatedly exposes the patient to bad memories so that they grow desensitized to them. The eye-movement component, they say, may be complete hooey. In fact, a dozen or so other studies do show similar rates of effectiveness between EMDR and cognitive behavioral therapy, but EMDR requires much less time recounting traumatic memories and more free association during recall. And, of course, more fun with crayons.

WE REPEATED THE entire four-part exercise with another event conjured from our time line, and this time I picked when Ennis told me about seeking out separated or divorced women. I drew a sarcophagus-shaped woman with a black coiled phone cord coming out of her ear and green excretions coming out of her mouth, black and red blotches along her lower torso. Distress scale: 6.5. Other drawings from that memory: jagged lines coming out of my head, dark cloud-like blotches. By the fourth tapping, I rated my distress level as 5.1. My future-self drawing was of two clasped hands and an eyeball, and the word *Touch*.

And so it went for that day and the next: picking out memories, drawing the feelings, and tapping. I alternated between events involving my husband and those involving Ennis. My future-self drawings felt like a gift: me surrounded by friends and family in a giant embrace, another of me standing under a starry sky with the word *Connected*. By lunch on the second day, I noticed a weird pattern, that my overall distress number was slightly higher around Ennis, and my pictures were disturbing too: lots of implements of violence, like icicles pointing my way, me lying facedown and immobile, a boot about to stomp on me. The future image after those drawings was a circle of loving arms around me, and the word was *Safe*.

During the lunch break, Chelsea and I headed to a nearby salad place to catch up a bit. She told me about her upcoming trip to work with refugees in South America. She was applying for jobs in trauma therapy. I told her how it was going for me during the workshop, and about Ennis.

"Interesting," she said, picking through her coleslaw. "You realize those feelings about Ennis are really about your husband, right?"

I looked at her.

"It's classic displacement," she said.

I ran through the drawings in my mind, the arrows to my heart, the stomping, the threats of confinement. Was she right? Could it be my husband who had made me feel unsafe? Of course. He was the source of my sense of peril, and not just now, post-separation, but through those last two years of our marriage, when I knew he was yearning for one woman and then another. Even when we tried to carry on as a couple, I didn't feel loved or safe or remotely secure. I had felt deeply wounded. Ennis had employed some of the literal tools of bondage, but I hadn't actually ever felt unsafe with him. I was mistaking the accoutrements for the feelings, but they weren't about Ennis at all. I had transposed his bag of tricks, with its gags and ropes and pita clamps—along with some of our unsettling conversations—and the *feeling* of being bound and imperiled. How could I not have seen it? I felt a pang of regret for the aspersions I'd been launching at Ennis. I'd been blaming him for what was actually the accumulation of years of feeling unheard in my marriage, along with the pain and humiliation of being left.

"Sometimes, EMDR therapy doesn't just process trauma," said Chelsea. "It reveals truth bombs."

———

THAT FINAL AFTERNOON of the workshop, we had one more memory retrieval. A totally new scene with my husband came to me, one that I hadn't even remembered to put on the time line. It was shortly after he had told me that he was again thrown by his feelings of love for another woman, that it hadn't developed into a physical relationship, and that he didn't know what he wanted to do. We'd gone to a party and she was there. He didn't know I knew who she was. It was unbearable to watch him with her. On that night I realized I would soon need to ask him a very hard question. Either let's fix us (please?) or let's just stop altogether because if this keeps happening I can't keep watching it. I had somehow blocked the memory from the earlier exercise. It was so much easier to focus on—and vilify—the man I'd known for only a few months than the one who'd spent 30 years growing roots deep into my heart.

Distress level: very high.

Drawing it out that afternoon, I conjured shackles around my wrists, another of me tied to train tracks. But it wasn't until after tapping and then starting to draw my final future-self image that I started sobbing.

I drew a woman in a raft, rowing herself down a gentle canyon river, and then the word hanging above it all: *Free*.

14

HIGH ISLAND

Warmth

Come back as a flower.
—STEVIE WONDER, FROM THE ALBUM
JOURNEY THROUGH THE SECRET LIFE OF PLANTS

Recognizing that my husband was the source of my feeling unsafe did not seem to be making me any calmer. It did, however, help me feel some constructive anger. I was now mad at him, less so at Ennis. And anger, according to science, is a necessary step toward moving on, as long as you're not burning the house down. Placing some blame on ex-partners and viewing them with a more jaundiced eye is known as "negative appraisal." In one study, people who more quickly associated negative words with their exes were less depressed and felt better faster. The study also found that people who'd been recently dumped—as opposed to those doing the dumping—tended to be slower to apply negative words, probably because they were still craving and missing their exes. And not surprisingly, they felt considerably worse.

Negative appraisal happens in your brain, but once again, it's the body that shows it. To test how unsafe I really felt, I visited Erica Hornstein, a postdoc in social psychology at UCLA, who once worked in the lab of the late University of Chicago loneliness guru John Cacioppo. She now mostly studies the opposite of loneliness: how social support helps us. She's particularly interested in how our support figures, like parents or partners, help provide a sense of safety when we feel threatened. In some experiments, she will show photos of a loved one (as well as pictures of a stranger) to a volunteer a few microseconds before delivering a shock to the wrist. During these experiments, she can measure "fear" by monitoring the volunteer's skin sweat and temperature via an electronic finger cuff. Her hypothesis is that we have a harder time learning to associate fear with loved ones. This makes sense evolutionarily—if mammal infants run away screaming from a parent, they won't survive. On the other hand, if Mom is truly a monster, her offspring do need to learn to fear her.

Hornstein offered to run me through a mash-up version, incorporating techniques from previous well-known experiments that use photos of an ex-partner, for example Helen Fisher's studies scanning brains of dumped lovers. So Hornstein set up a comparison: she would give me electrical shocks while I viewed pictures of my father (my support figure), my ex, and a random old white guy from the internet. The central question she was asking: Would I associate fear with my ex, or was he, as a past support figure, still filling the role of safety? I wasn't so excited about the shock part—47.5 volts delivered to my right wrist. Walking down the hallway to the featureless lab chamber was like entering torture Room 101 from *1984*, the place where your worst fears make you forget love.

In the first trial of the experiment, Hornstein showed me two pictures of each figure, one of which was paired a couple of microseconds later with a shock, repeated several times. Up flashed a photo of my elderly, white-bearded, kind-eyed father in front of a waterfall: no shock. Photo of Dad by a river: shock! My husband on a rooftop at a Fourth-of-July party: no shock. Him at an outdoor café: shock! Same story with the random internet guy. The wrist shocks felt like an unpleasant pinchy buzz. Then, during a short break, Hornstein played two minutes of a "neutral" video tour of a 747 aircraft. Then I watched as she turned up the shock dial, *Princess Bride*–style, and showed me the photos again. Only this time, it was a trick; there were no shocks. This iteration would supposedly allow my brain to decouple the fear from the photos. The idea was that after several rounds of this, my nervous system would get the message that these men were all safe now.

Later in the day, I stopped by Hornstein's office for the results, which she had nicely graphed out. As she expected based on her previous work, I was not able to learn to associate fear with my father. Even though I would receive a shock microseconds after I viewed one of his photos, my skin stayed dry and cool, indicating I was not afraid of his image (I was still reacting in a fearful way to the shock, but my body wasn't preparing for the pain ahead of time when looking at his kind face). I did, however, feel fear and anticipate pain when viewing the photos of the stranger (also expected) and my husband (interesting!). In fact, during the "deconditioning" trial, it took longer for me to extinguish the fear from my ex-partner than it did from the stranger. It wasn't that any photo of him made me freak out, because the neutral nonshock pictures of him, designed to be a helpful baseline, showed no big rise in dread (although they did show a little rise, which, as Hornstein pointed out, is a good reason not to follow one's ex on social media or cross paths with him or her too much if you want to feel safe). When the shocks subsided, it took longest for me to trust the photo of my ex. "He's either already more threatening, or you're sort of more ready to view him that way," said Hornstein. The message, at least as far as my nervous system was concerned, was pretty clear: when you've been hurt by someone you love, your body remembers.

AT THE 50TH birthday party of a friend in my neighborhood, her husband of 25 years made a heartfelt toast. "You are the yin to my yang," he said. "I can't wait to be with you for the next 25 years," and he looked like he meant it. I tried to appear cheerful. Then I bolted for the back deck, where a guy named Sandro was passing around a large bong. I knew Sandro by reputation. He lived across the river, in Virginia. A jazz bassist, he grew his own cannabis plants and sold the offerings to friends and neighbors. I liked his looks. He was a man of appealing trapezoids, from his muscular calves to the solid cabinetry of his trunk.

We both liked rivers, and camping. He offered to take me and some friends to camp overnight on an island in the middle of the Potomac. No one was sure if the island was private, or even which state it was in, but he'd been there before. So one Sunday evening when I didn't have the kids, Sandro and I set out. The friends had all bailed, smartly so, because it was late fall, and freezing. We paddled out in the last light, landed, and built a fire in a waiting firepit. He had stashed a tent there on an earlier expedition, and we set it up. We stayed up late, talking by the fire, eating spoonfuls of peanut butter out of a jar.

Awkwardly, we splayed our sleeping bags close together in the large tent. It was seriously cold. Inside, our down-wrapped bodies touched. Then they moved closer against each other, and closer. He smelled like smoke and loam and peanuts. Gradually, the sleeping bags opened. Sandro's body burned like a furnace. Where I was frenetic and twitchy, he was solid-mass geometry. With him, I stopped fluttering. I grew warm and steady.

My body might not have wanted to be near my husband, but it sure wanted someone. It wanted to feel safe, to calm down, to feel loved. Like all mammals, it wanted to be warm. It wanted, when stressed like a prairie vole, to huddle.

Many psychologists, my own therapist included, are wary of people seeking other relationships too soon after heartbreak. But I was surprised to experience how strongly the body wants what it wants. For years, decades, I had instead let my head determine my actions. I was like James Joyce's character Mr. Duffy, who "lived at a little distance from his body." Marriage had vitiated my sensuality. Now my body was asserting itself, churlish and girlish.

Maybe my physical craving was my need for self-substantiation, to prove I existed, and that I was made of matter even if I seemed to be bodily dissolving. Or maybe it was just hormones. Men's bodies start manufacturing more testosterone around the time of divorce, according to a 1998 study of 4,500 American veterans. Evolutionarily, this makes sense. On their own, they might need to fight, or at least take bold, assertive action to find a new mate. Hookup apps are crowded with freshly sprung husbands. They are horny, and they are lonely. Divorced women also show elevated levels of testosterone, and it influences our sex drive and our competitiveness.

According to my medical labs, I was also still producing plenty of estrogen. For this, I thanked the universe. Being left by a partner at 50 already felt like twisted karmic punishment. Menopause—along with its reported dip in libido—might be arriving any moment, but it wasn't here yet. A recent study from University College London looked at health records from 2,900 women in their late 40s and early 50s, finding that those who had sex at least weekly were more likely to delay the onset of menopause than women who had sex less than once a month. Some—but not all—of my friends over 50 report having great sex, and I wanted to surf my hormones while I still had them.

For months through the long winter, we dashed across the bridge to each other's houses and adhered our incongruent bodies together. A kleptotherm, I stole his heat. He warmed smooth round rocks on his wood-burning stove and placed them on my bare skin. He fed me stews. I noticed that after a night under Sandro's furry, weighted arms, I would wake well rested and calm. My morning glucose levels would miraculously drop.

For his birthday I gave Sandro a sheepskin rug. We would splay across it, never talking much. The opposite of Ennis, Sandro touched me gingerly, as if I were a baby bird, and I liked that, too. "How do you reteach a thing its loveliness?" asked the poet Galway Kinnell. "To put a hand on its brow / . . . and retell it in words and in touch / it is lovely."

I'VE TALKED ABOUT the physical pain of separation and how those aversive signals—the pain part—drive us not to separate. But apparently that isn't enough to control the attachment switch. There's also a powerful flip side, which is the pleasure of connection that impels us together. As mammals evolved, social pain piggybacked on the physical pain networks in the brain. Similarly, the social attachment system appears to have piggybacked on the brain's opioid system, part of the pain network. The areas of the brain that light up during partner separation—and during reunion—are notably rich in opioid receptors.

How does the brain know we are physically close to someone? Our skin is our largest sensory organ. Sometimes it is described as our social organ. We mammals have specialized nerve fibers under the skin called cutaneous C afferents. These nerves love to be caressed at slow speeds, according to science (in case we didn't already know). When we are touched in this way, our breathing slows, our blood pressure drops, and our muscles relax. Psychologists have measured the way having friends or loved ones holding your hand can help mitigate pain in patients undergoing medical procedures. But recently they have begun to look at touch and the easing of social pain.

In the mean-girl Cyberball game, players pass the ball to everyone but you, creating feelings of rejection and ostracism. Touch researchers at University College London made 84 volunteers play the game. Then the researchers blindfolded the players, and stroked half of them slowly (on the forearm) with a soft brush ("blush brush No. 7," to be exact). The other half got a fast-brush treatment (same brush), administered at 18 centimeters per second instead of 3. After that, the volunteers filled out a questionnaire measuring their threat level. Both groups still felt hurt, but the slow-touch group reported feeling significantly less distress and threat. Of course, all good parents understand the benefits of comfort strokes, along with kissing boo-boos and even a good tickle session, but scientists have a fancy name for it: embodied social support. Adults need it too.

Some of our skin nerves also monitor temperature, and these send signals to our brain stems to tell us to start shivering, or to burn brown fat, or to find Mom or, in my case, the local bass player. This is a deeply hardwired urge.

Mammals must maintain a very narrow range of body temperature in order to survive. It takes a lot of energy to be warm. One researcher measured how group huddling allows newborn porcupines to thrive in colder temperatures (somehow, they work out all the poking).

When we're warm, our bodies release more natural opioids, which may be one of the reasons frequent sauna use in Finland (and aerobic exercise) is associated with lower rates of depression, stress, and stress-related diseases like heart attacks.

Studies show that when people hold a hot beverage, they subconsciously behave with more "warmth" and generosity toward others. When they sit in a warm lab room versus a cold room, they report feeling closer to the experimenter. On the other hand, when subjects are asked to recall an experience in which they were rejected, they estimate the room temperature to be colder than people recalling inclusive events. The rejectees also report a greater desire for warm food and drinks. When the participants are colder, they report feeling lonelier. Notably, researchers have suggested that "experiencing the warmth of an object could reduce the negative experience of social exclusion."

To see if lonely people routinely (and subconsciously) "substitute" warm objects or behaviors for missing relationships, psychologists at Yale University asked 100 students and older community members to fill out the UCLA loneliness scale. They also inquired about their bathing habits. They found a significant positive association between ratings of lone-liness and the frequency, duration, and temperature of showers and baths, such that about a third of the variation in bathing could be explained by loneliness scores alone. In the journal *Emotion*, the researchers, John Bargh and Idit Shalev, wrote: "Together, these findings suggest that phys-ical and social warmth are to some extent substitutable in daily life."

I was becoming a master substituter. On the nights I wasn't with San-dro, I would linger in hot baths. I often went to bed with a floppy red hot-water bottle, which made me feel ridiculous and geriatric, but not enough to override the pleasure of doing it. In the months after my breakup, I was rarely without a thermos of tea. It was practically, embarrassingly, my security blanket, what child psychologists call a "transitional object." I may not have had a partner, but I had a thermos. If, jet-lagged and pre-occupied, I accidentally left it behind (which I did twice), I was bereft. And cold. And, apparently, more lonely. One doesn't typically read this in breakup manuals, but now the science is in: *seek heat.*

NOT ONLY DOES pleasant physical contact release opioids into our brains, it also releases oxytocin, which in turn activates reward neuro-chemicals like dopamine and serotonin (and when dogs and their owners reunite, both animals release oxytocin, especially dogs when the reunion involves petting). The more readily these flow, the less we release stress hormones like cortisol, and cortisol plays a direct role in regulating blood sugar. Sandro was my heat source, orgasmatron, and insulin port all in one.

In a series of studies, when researchers gave oxytocin to rats, they didn't mind being burned with heat probes as much as rats who didn't get the oxytocin. And in human experiments, students in new romantic rela-tionships showed less physical stress (as measured by heart-rate variabil-ity) when viewing disturbing videos than single students. Those healthful benefits of oxytocin—once thought mainly to facilitate bonding—may not be incidental; they may be its original purpose. "The real power of love is through the actions on these really primitive systems that are there for the defense of cells and whole organisms, like inflammation," pioneering oxytocin researcher Sue Carter, now an emeritus professor at the Kinsey Institute at Indiana University, recently said. When we are buffeted by fear and anxiety, oxytocin is a natural fire extinguisher. Divorce may be an inflammation story; happy sex is an anti-inflammation one. If love is a physical and emotional analgesic, it makes sense that we seek it when we are hurting.

Lovers mysteriously help "coregulate" each other in other ways, too, which is another reason we feel their absence so acutely. As the neuroscientist David Eagleman puts it, "People you love become a part of you—not just metaphorically, but physically. Your brain refashions itself around the expectation of their presence." In a study published in 2020, researchers in Finland inserted 10 couples in a two-person scanner, face-to-face, while they took turns gently tapping each other's lower lips (must be a Finnish thing). Regardless of who was doing the tapping and who was doing the feeling, the couple's brains synced up (albeit with slightly different intensity) in the sensory and motor parts in a sort of Vulcan mind meld. One person's brain was "resonating" in the other's, concluded the authors. Sleeping practically on top of someone, your heart rates and respiration speeds begin to synchronize. So do cortisol levels, which, remarkably, tend to align on mornings and evenings in couples who live together. This is a good reason to find a relaxed partner if you tend to be agitated, or to avoid a depressed partner if you enjoy a fast pace.

Perhaps this is all just a fancy rationalization for why I wanted to be dating. I was aware, of course, of the received wisdom about the pitfalls of rebounding too fast. I listened to one podcast advising divorced people to wait years before dating again. One oft-repeated formula states that for every year of marriage you should wait six months before embarking on a new relationship. That would put me on the 12-year plan. Sorry, no. I might not be eating solid food by then. Many popular articles on the subject—as well as many psychologists—issue bromides about learning to love yourself first. They warn that jumping into new relationships is misguided at best, as well as a problematic distraction from confronting your feelings.

Pema Chodron, the Buddhist nun who left her first husband and was left by her second, is wise on the subject of suffering. I love so much of what she says, such as the necessity to turn "hot" loneliness into "cool" loneliness, a state of less desperation, more equanimity. But I'm less sure about her prescription for getting there. "Heartache is not something we choose to invite in," she writes. "It's restless and pregnant and hot with the desire to escape and find something or someone to keep us company." Best, she says, to exercise self-control and thus come to know ourselves better.

Until I'm ready for enlightenment, though, I'd rather have the company. It may hold Buddhist cred, but the Step-Away-from-the-Cute-Boy wisdom has little basis in science. In 2014, researchers at Queens College, City University of New York, decided to study rebounders. In two different studies, they asked many questions of over 300 volunteers ranging in age from 18 to 49, all of whom had recently (on average 7 months earlier) experienced a breakup from a relationship that had lasted, on average, 11 months. The subject group was racially diverse but overwhelmingly heterosexual. The researchers, Claudia Brumbaugh and Chris Fraley, found that those who had quickly entered into a rebound relationship tended to report feeling more insecure in general than the ones who were still single. Despite this, they reported more confidence in their desirability, suggesting that the act of rebounding was lifting their damaged self-esteem.

The quicker rebounders were also less likely to report lingering romantic feelings for their ex, and they reported higher overall well-being. "This suggests that having a new partner may effectively serve the purpose of allowing people to more quickly get over their ex, even when the breakup occurred recently," the authors wrote. "In sum, people who could be described as rebounding tended to have better personal psychological outcomes and valued their new partner more."

As to the distraction-from-feelings argument, Brumbaugh and Fraley pointed out that this is called coping, and it's not always a bad thing. "People often experience despair and loneliness following the end of a relationship," they wrote. "Dating a new person may provide a way to divert attention from these negative emotions." I was aware of the significant risk of depression following a breakup (remember, it's 23 percent). Once you're deep into a serious bout of depression, it can be very difficult to climb out. With each successive bout, you're more likely to experience another. I didn't want to run away from sadness, but I didn't want to wallow in it to the point of dysfunction. It wasn't just the threat of depression sabotaging my work and my parenting. To succumb to that would have been allowing the breakup to take an ever bigger piece of me. I had more fight in me than that. And my C afferents liked it too.

15

EXCUSE MY PILOERECTION

The Science of Awe

What are men to rocks and mountains?
—JANE AUSTEN, *PRIDE AND PREJUDICE*

On Valentine's Day, the kids and I went out for pizza, something I rarely ate since learning about my deficient pancreas. We talked about the river trip. They would be joining me for sections, and launch dates were coming together. It was a happier conversation than the one we'd had the previous evening over dinner at home, when they asked me why I'd just had such a long meeting with their father. Right. I smoothed out my napkin and launched in.

"Dad and I have been having discussions about how to make our separation more permanent," I said.

"You mean, divorce?" said my daughter. We'd been eating burgers and slaw, sitting at one end of the dining room table.

"Yes. I'm sorry. I know it's not what you were hoping."

"We didn't think you were getting back together," said my son. "You don't have to be sorry. We don't think it was your fault."

This was news to me. They'd never asked about the reasons for the separation and we'd never discussed it.

"I just meant I'm sorry that it's happening," I said. "It wasn't my first choice for how to do things, that's true, and I'm sad, and it's hard." I started tearing up.

"We're here for you, Mom," said my son. I reached over and squeezed their hands.

"Do you think you'll remarry?" asked my daughter.

"Maybe. I'd like to be in love and find a partner someday."

"Do you think Dad will remarry?" said my daughter. "I can see it in like 20 years."

That made me laugh. "I think it will be much sooner than that!"

Back home from pizza and settling in for the evening, I checked my blood sugar, which I do by poking myself in the finger. The meter flashed a number I'd never seen before: 359. Usually my glucose level after eating is half that. I called a friend who has diabetes.

"What do I do?" I asked.

"Go to the emergency room," she said. "Don't go alone." She was worried I'd get spaced out. I was already feeling a little weird, like I was underwater.

It was 9 p.m. I knocked on my son's door. I was glad he had a driver's license. The hospital was close. Once there, we poked at our phones during the long wait to see a doctor. My son tried to reach his father, but he was in a bar on a Valentine's date and didn't respond. I texted Sandro, who said he was "very baked" at a friend's house but would check in later. Okay, then. The landscape of deliverance was stark. I was leaning on my kids now, and that felt upside down but also necessary. My son was calm, cheerful, concerned. He was good in a crisis, I was learning. I felt proud of him, and lucky.

The doctor was unimpressed. My blood glucose was by now declining on its own. It was nothing compared to most of the diabetics he saw in the ER. He told me not to come back unless my level registered in the 500s and I was going blind or numb. "Don't eat so much pizza," he advised, and dismissed us into the late night.

I was fine, but shaken. On Valentine's Day, my body was sending a clear message; it was still attached to my heart, and all was not well. Message received.

I BROKE UP with Sandro. I couldn't figure out what we were. Whatever it was, it didn't feel like enough. Then we got back together. Then we broke up. I was restless. I wanted to calm down and heal my body, but that was hard to square with the feelings of impatience and my need to spin around, to shake the trees and make some fruit fall down. I wanted the universe to give me answers and to give me comforts, ideally at the same time. I went hiking with my friend Eric, who was writing a book about philosophers. He quoted the French political and religious thinker Simone Weil to me. "Above all, our thoughts should be empty, waiting, not seeking anything, but ready to receive." That seemed impossible.

Eric explained that Weil, who was something of a mystic and died of tuberculosis and starvation at the age of 34, believed that much suffering in the world occurs because we attempt to "eat beauty rather than see it." What she meant, he explained, was that when we pursue something, when we try to possess it and make it serve us, it becomes about our needs and our ego, and then we can't really love what we are seeing.

Eric was gently implying I had a consumer mindset. I wanted to eat up every data-driven heartbreak solution. Maybe I should have spent more time reading philosophers and less time reading neuroscience studies. Although, ultimately, they point to similar truths. Expect to suffer. You can't run away from pain for long. You must feel it and then you must wait. And yes, beauty could blow open the possibilities of the universe while also making you less focused on yourself. As Emerson said, "The question of Beauty takes us out of surfaces, to thinking of the foundation of things."

I didn't want to just consume beauty. I wanted it to consume me. The harder I looked into the science, the more I learned that there was one singular emotion that conjures this experience: awe. Paula Williams's work on beauty and aesthetic chill was just one small piece of it. I'd been watching sunsets for months, walking through forests, backpacking through blizzards, inhaling the smells of the woods and of men. I did these things because they felt good, but I can't deny there was an agenda (I'm sorry, Simone). I hoped to put my problems into perspective, to calm down, to gain understanding about what had happened. Awe, I was learning, helps on all three fronts.

Consider perspective. Most of us can relate to the experience of looking up at the night sky and feeling like a small piece of lint in the universe. We feel we are a part of something larger than ourselves. We feel connected to the natural phenomenon, but also to the human family. Studies have shown that after viewing images of epic nature (as opposed to images of something more mundane like a shopping mall), people behave in more generous ways. They give away more money to other players in lab games; they fold more paper airplanes for earthquake victims; they volunteer more time to help others. By prompting us to be better humans, the data was suggesting, experiencing nature is good for civilization.

To distinguish the effects of awe from other positive emotions, researchers at the University of California Berkeley asked volunteers to recall experiences that made them feel good, such as attending parties, being in nature, and accomplishing goals, then list the emotions, and then rank a series of words or phrases related to self-appraisal, such as "meaningful," "challenging," "felt connected with my personal values," and so on. When they recalled feeling awe, they ranked the phrases "felt small or insignificant," "felt connected to the world around me," "unaware of day-to-day concerns," and "felt presence of something greater than self" nearly twice as high as when they experienced other happy emotions. In another study, the same researchers took volunteers to stand for one minute either beneath a full-size replica of a *Tyrannosaurus rex* skeleton or in an empty hallway. When the volunteers were then asked to list 20 statements to answer the question "Who am I?" the T-rex group used three times as many "oceanic" and universal descriptors—such as feeling they belong to a community—and significantly fewer statements defined as individualistic.

Awe may also make us feel literally smaller than we actually are. This shift in body perception was suggested by a pairing of studies by researchers in the Netherlands published in 2016 and 2019. In the first study, participants watched videos of either vast nature (waterfalls, oceans, mountaintops, etc.), cute pets (a positive-feeling control), or views from a car on a highway (a neutral control). The people who reported feeling the most awe described their physical bodies as being slightly smaller than they actually were, a subtle difference compared to the people who did not report being awestruck. In a field study from 2017 in California, volunteers represented their bodies as being 33 percent smaller after looking at majestic views in Yosemite than after looking at a street by Fisherman's Wharf in San Francisco. Notably, however, feeling literally smaller did not lower their self-esteem; it just made other people and landscapes seem relatively more significant. This is a literal representation of what researchers mean by the phrase *healthy perspective.*

As far as calming down, the literature points to something counterintuitive. When you are struck by amazement, you might think your heart rate and respiration increase. But when you are experiencing "positive" awe, as opposed to threatening awe (like a tornado), your parasympathetic nervous system—the calming one—engages. Craig Anderson, a psychologist at the University of California San Francisco, posits that the body becomes more still in order to absorb the experience and make sense of it. Even our facial expressions slow down and freeze, he said; our jaws drop and our eyebrows rise.

A key component of awe is surprise, the sense that we are seeing something unusual that we don't fully comprehend. Anderson described to me a funny video that had been going around the internet of a baby crying in his car seat. The car suddenly goes through a dark tunnel, and the baby stops fussing and opens his mouth. He looks completely, happily, gobsmacked.

If people's faces and heart rates change in the presence of awe, what about their immune systems? Anderson was part of a team that measured how cells look in people who report frequently experiencing awe. They asked college students to recall how many times that day they felt positive and negative emotions from a long list. They also took samples from each volunteer to measure interleukin-6 (IL-6), a cytokine and one of the same markers of inflammation that I'd been checking in myself at various time points since the breakup. The emotion most strongly linked to a healthful change in IL-6 was, you guessed it, awe. Shame, on the other hand, has been shown in other studies to drive the opposite change in IL-6. It matters because high, persistent levels of IL-6 are associated with diabetes, cardiovascular disease, and depression.

Why would experiencing awe be linked to these changes? The authors had two theories. Either people with naturally low inflammation are more likely to find awe because they have the health and energy to get out there and experience beauty, or the regular encounters with awe are settling them down and infusing them with positive feeling. Or, perhaps, it's a little bit of both, supporting Paula Williams's conviction that we can learn to make ourselves more awe-prone, and eventually healthier.

"I think of awe as an emotion that helps us explore," said Anderson. "It helps us be curious." In this way, awe facilitates the third bullet point on my heartbreak-recovery agenda after perspective and calm: making sense of what had happened and finding meaning in it.

Melanie Rudd, a marketing scholar at the Bauer College of Business at the University of Houston, and her colleagues decided to test Anderson's curiosity hypothesis in the field. To provide some serious awe, they set up shop in the Swiss Alps, where they persuaded 162 tram riders to complete a study in exchange for trail mix. Research assistants (who were blind to the hypothesis) approached the tram riders at one of two field stations: either at the top of the mountain, where the "awe" quotient was extravagant, or at the bottom, near the parking lot. They asked the tourists to take a short survey, then offered them the opportunity to grab some premade trail mix or to mix their own, both options involving dried apricots, cashews, raisins, and hazelnuts. Then the researchers offered an educational brochure with information about the region in case they wanted to learn more. Rudd found that the tourists surrounded by Alp views were 40 percent more likely to make their own snack creation and 36 percent more likely than their parking lot peers to accept the educational brochure (everyone was eventually going to the top). To her, the snack scenario confirmed the idea that awe renders us both more "time-affluent" (another study of hers suggested awe makes us feel like time slows down) and more creative. We want to make things! Put ideas together! The brochure-taking, she felt, indicated an increased openness to learning things and a desire to make sense of information.

The thing about awe, as so many of the experts point out, is that we don't readily comprehend what we're looking at. We seek to update our mental maps to accommodate this new amazing spectacle. Normally our brains group things into easy categories in order to get through the demands of the day. We rely on those stored categories of knowledge as we process new information. Awe doesn't let us get away with that. It makes us stop and pay attention.

Michelle Shiota, a psychologist at Arizona State University, says awe encourages us to absorb new data points and even change our minds about something (the sun can be orange and even vanish altogether during an eclipse; the sky can be purple; hail as big as gumdrops can fall on your head; a baby comes out of your body and suddenly you are filled with love for a being you've never seen before; the strains of an oboe can make your entire body vibrate). Perhaps, then, awe can actually provide an opportunity to reassess ourselves.

"It suggests that if you have a schema of yourself, like a relationship that needs to be modified, it's plausible that awe could facilitate adjustment," Shiota said. It's tricky to measure, but she believes it's possible to show that a person's self-concept changes more rapidly when she finds herself in an awe state. This might be part of why, for example, wilderness-therapy programs, corporate rafting retreats, and extraordinary travel experiences can bring about big changes in how we view ourselves, our team, or the world. It might be why, as Paula Williams has suggested, people who are awe-prone can reorient themselves to a happier story.

We shouldn't necessarily assume, warned Shiota, that all awe experiences will change us for the better. Through the ages, powerful rulers have elicited awe through giant pyramids or ermine robes or divine right or massive rallies, convincing us to submit to their authority. Many humanists see a vast and overpowering event like a pandemic—an event that fits the definition of something truly (negatively) awe inspiring—and warn that it could shift us either toward accepting despotism or toward aspiring to greater equality, human connection, and environmental stewardship.

But with the right support, the right guidance, or the serendipity of good timing, it's possible that experiencing awe can help us swerve toward becoming more helpful, purposeful, and resilient, or toward any other goal. "The question I'm interested in," said Shiota, "is whether awe provides an opportunity to reshape our understanding of the world in ways that we choose."

It was, to me, a profound and elegant idea: awe could provide a rare window of malleability.

It was time to bust through that window and tumble into a river.

16

SPLIT MOUNTAIN

When it hurts, we return to
the banks of certain rivers.
—CZESLAW MIŁOSZ, "I SLEEP A LOT"

I.

Starting a wilderness trip at a giant wall of concrete may seem odd, but
this is the American West. Many such giant walls—dams—punctuate
the mightiest rivers. At these points, they chokehold the water, greed-
ily taming it into large lakes to keep irrigation reliable throughout the
dry season. There are unfortunate consequences: the dams disrupt the
natural hydrograph or flood pattern, causing erosion, destroying habitat, impeding fish migration, and making it easier to lose precious water
through evaporation.

The upside, though, can be a pretty nice manufactured trout fish-
ery in the clear, cool, flowing waters below the dam. The morning we

dropped two inflatable kayaks and one overstuffed raft into the Green River, a pack of anglers surrounded us, winching their sleek drift boats off trailers at the launch ramp. Most were on day trips and would take out long before us. We'd be out here for three days paddling 30 miles through a red canyon called Flaming Gorge that opens up to the wide grassy benches of Brown's Park in far-western Colorado.

It was an exciting place for me to start. I'd never run this section of the river. And it was a relief to finally launch, after so much planning, so many metaphorical and physical things to tie up and tie down. I liked that this stretch was new. For six years, we'd lived a half day's drive from here, in Steamboat Springs, while my husband was working on river conservation and I was freelancing and writing a graduate thesis. We'd run rivers near here. Being in this new stretch was staking a claim on rivers as my tradition, not just our shared one. And I was pleased to be passing it along to our son, Ben, who, at 16, was joining me before heading up to northern Wyoming to spend five weeks at a backpacking program. My younger half brother Berkeley was here, too. An accomplished rower, he'd spent years on river trips with our dad, then with me and my husband and the kids. My friend Doug and his two younger boys also joined us.

A few days earlier, back in DC, my husband had come by to talk over some final points in the divorce agreement. I asked his advice about the fishing. Which rods, which leaders, which flies? Our family gear was still mostly jumbled together and I was sorting it out. I had my own rod, but I'd used it mostly on small Montana rivers, and the Green would likely demand something sturdier. As with the kids and the dog and the pickup truck, we'd decided to share custody of our raft and most of the essential river gear, which we keep stored in Berkeley's garage in Colorado in exchange for use of it. Ironically, perhaps, the raft and the oars had been a wedding present, so it was communal property in the purest sense. He was helpful and matter-of-fact. He even tied a special leader onto the fishing line.

"It feels weird doing a river trip without you," I said, zipping the rod cases back up.

"Yeah, that's why I want to help Ben with the rods," he said, "and Berkeley knows the boat."

"That's not what I mean. I mean our divorce. It still feels strange and wrong. It's just such a nuclear option." It was too late to be dragging this out again, but it wanted to be said among the jumble of shared waders, duffels, and sad dry flies.

He nodded and shrugged and got into his car in the rain. He was driving to his girlfriend's house for the weekend.

DESPITE HIS EFFORTS, the fishing tackle wasn't quite right for the season. Doug had a beautiful cast, and kept at it all day from the raft and from shore, but the fish would have none of it. Ben just wanted to row. He was now six foot two, and he had somehow acquired arm muscles. The raft was heavy in the current, but he could move it. Berkeley sat on a pile of gear bags giving him some instruction and also trying, without success, to catch fish.

The younger boys were paddling a double ducky—an inflatable canoe—and I paddled my packraft, a delightful little seven-and-a-half-foot boat that my husband had given me for a recent birthday. Perhaps, in retrospect, a parting gift. I hadn't used it much. But now the boat and I had long days ahead of us, and I wanted to make sure it was comfortable and airtight. Weighing only six pounds empty and made from some miracle laminated nylon in a shade of blue found nowhere in nature, it felt flimsier than it was. Kitting it out, I'd spent part of the second morning on the river blowing big gulps of air into my sleeping pad to just the right proportions to use as a footrest.

"Careful, Mom, you're going to pass out," said Ben, which cracked me up so I had to start over. When did he become so arch and cool and large limbed and capable? He'd driven the truck the entire eight hours here from Boulder. This is a boy who loves a stick shift. It was hot out. We'd had to keep pouring water into a portable swamp cooler perched on some luggage in the cab and occasionally resting bags of ice on our heads.

He'd dialed up "vintage vinyl" on his phone and we listened to the Clash, Queen, Meat Loaf, and Joan Jett as we crossed the rolling shrub steppes of the Continental Divide. Now, on the boat, he was wearing mirrored aviator sunglasses and resting his arms on the big oars. The water was lively but gentle, glinting in the sun and lapping up against the boats as we bobbed along under a blue sky.

I was slowly shedding the city, shedding the divorce. I looked at Ben and Berkeley in the carefully packed boat, floating like a slab of blue steak on the water, and thought, *Look at us! Who needs a husband?* We were paddling well, eating well (no trout, but still), playing games around the campfire at night. The moon was nearly full. My boat was not leaking. I was sleeping.

Three days on the river, and we were on schedule for experiencing the so-called three-day effect. Nearly a year earlier, I'd spent four days on a different stretch of this same river south of here with a group of veterans for a podcast I was making. We'd been following different groups into the wilderness and watching and sometimes measuring how three days outside changed their brains. On that trip, Dave Strayer's doctoral students from the neuropsych lab at the University of Utah fitted the veterans' heads (and mine) with EEG units to record brain waves at different time points. Months later, after the data was crunched, the lab found not only that the participants reported better, happier moods after three days outside but that our brain waves appeared to suggest patterns associated more with be-here-be-now mindfulness.

Strayer believes several things happen to our brains during a multiday nature immersion that make us better humans. One is that the sensory parts of the brain wake up, and our usually dominant attention network quiets down, resulting in some new and expansive neuronal connections that lead to creativity. Brain imaging by his lab and elsewhere suggests that immersive nature-time partly quiets the frontal cortex, where we solve problems, manage our to-do lists, and ruminate about ourselves. Outside, he says, we are essentially resting these overworked neurons, so when we return to problem-solving, we are suddenly sharper, at least for a time. He and colleagues tested backpackers before and after a three-night trip. The subjects performed 47 percent better on word problems at the end of the hike.

There aren't many psych studies looking at river-running, specifically. But one, of canoeists in the Boundary Waters between the US and Canada, found, similar to Strayer, a 50 percent improvement in creativity after several days in the wilderness. (Lest you think the boost might be a practice effect, a control group of subjects who stayed at home and took the tests several days apart showed no meaningful change in scores. Nor did it appear to be a vacation effect: a different study compared backpackers to urban sightseers, and found increased cognitive performance only in the backpackers.) But what were the implications for an emotional reboot? Another intriguing study looked at inner-city teens and veterans with PTSD who went rafting and camping for several days in California's Sierras. The researchers found significant increases in well-being one week after the trip compared to before the trip, and the emotion most clearly tied to that improvement was—perhaps you guessed it—awe.

There are other factors at work. Everyone knows food tastes better outside, probably because we're freaking hungry, but another theory is that we're just taking deeper, calmer breaths and letting the "rest and digest" branch of our nervous system happily do its thing, pumping out optimal enzymes, taking its time. But why do we sleep better? The lack of urban noise, electric lights, and not much else to do at night all play a role. A study of campers out for a long weekend showed that sleeping outside reset their circadian rhythms to the point where they fell asleep earlier and averaged an hour more sleep per night, benefits that lasted for some time once they returned home. Natural daylight is 100 to 1,000 times brighter than indoor lighting, and the darkness is darker, too, triggering the full release of hormones that tell our brains how sleepy or wakeful to be.

Behold the three-day effect. It's a combination of being less under the

influence of built-world problems and more under the influence of the
sun, river, and sky. As veteran and river runner Stacy Bare had explained
it to me, he believes our reality exists in a kind of suspension of time, the
boundaries of which extend only about 72 hours. Most of our working
memory lasts about that long. So by the third day outside, our new frame
of reference is what's right here. "The three-day effect turns on like clock-
work," he said.

The educator Robert Greenway was a founder of the ecopsychology
movement in the 1970s. In a study of 700 wilderness-therapy participants,
he found 82 percent of them experienced a dramatic shift in dreams after
three days outside, from dreams dominated by urban settings to dreams
about the group or the wilderness. "Our culture is only about four days
deep," he wrote.

My reality was now tangled fishing lines, my seaworthy vessel, the
positive masculine energy around me (there was such a thing!), mos-
quitoes at dusk, and some extraordinary country. We were traveling
through the impressive red canyon, then a verdant valley, and lying
behind that, what Wallace Stegner called "broken badlands" in his
account of Major John Wesley Powell's expeditions down this river in
1869 and 1871. I was reading Powell's own journals and other write-ups
of those descents. His first expedition was made up of scofflaws and
misfits. None had ever run a rapid before, and the heavy, keel-bottomed
wooden boats they chose were spectacularly ill-suited for the trip. Seven
of the ten men were Civil War veterans, physically and emotionally battered and desperate for paying work. Powell himself had one arm, the
other lost to a musket ball at the Battle of Shiloh. He had to be literally
tied to a dining chair atop the small, slippery deck. Powell's brother,
Walter, suffered from PTSD. Stegner describes the major as surly and
unpredictable, not exactly ideal attributes for leading a long and dangerous expedition.

The party's objective was twofold: to survey a vast, inaccessible territory, including a thousand miles of the Green and Colorado River sys-

tems wholly unknown to colonizers and bureaucrats (the Utes had been here for at least hundreds of years), and for these restless men themselves to do something other than farming or fighting. Maybe they'd find some gold to boot. They couldn't have anticipated the dire rapids, the starvation rations, the impossibility of walking out. The river flowed fast in one direction; each day would take them deeper into desolation. Starting a bit upstream from here in late May, the river raged with snowmelt.

"The cañon is much narrower than any we have seen," Powell wrote of Flaming Gorge, which he named. "With difficulty we manage our boats. They spin about from side to side, and we know not where we are going, and find it impossible to keep them headed down the stream."

Between my new 72-hour memory reset and the ghosts of Powell, being out here was like riding a time machine. I was flowing away from the broken badlands of my marriage. I was getting sucked out of my recent past to emerge somewhere much, much more pleasant, more physical, more sensory, more playful, and maybe, I hoped, more in tune with an earlier version of myself.

II.

For the second section of river, it was my daughter's turn. We'd be running Lodore Canyon, which churns up the biggest rapids of the whole river. I didn't draw a private permit through the competitive lottery for this section, so we joined a commercial group that would let me paddle a boat, and they'd even cook for us. I liked the company, called OARS, which had outfitted one of the earlier reporting trips I'd taken with the veterans. This was considered a family trip, so there was a group of kids around Annabel's age, 14. Teens and soda: she was psyched.

This 53-mile section of the Green River starts unimpressively near Vernal, Utah, in a flat, sage-strewn reach. But within a mile and a half, it enters an impressive rock chasm known as the Gates of Lodore. The river no longer flows clear here, but rather faded green, narrowing and quickening as high walls rise up on either side. It's like entering a birth canal.

You're not sure how you'll come out the other side, but you suspect you might be different.

That first morning, Annabel chose to paddle with me in a two-person ducky rather than sit in a raft. I was thrilled to share it with her, but I also felt like an imposter mom pretending to hold it all together. Drifting in the boat for some moments of calm water, I was still zinging from the surreal emotional calisthenics of the previous days. After the Flaming Gorge leg, I'd left the river for a couple of weeks in order to accommodate the kids' schedules, complete some paying work, and, most importantly, finalize the divorce settlement agreement. I'd wanted it done before launching into this next three-and-a-half-week heart of the river journey, meant in every way to be a liminal reach between my old life and my new one. In the past couple of days, we'd scrambled to finalize details with our respective attorneys, transfer funds, close bank accounts, open new ones, proofread documents, all while I was running errands for last-minute expedition items. It had come down to the last possible moment. An overnight packet of documents with the final agreement arrived at my sister-in-law's house in Boulder (where I was staying) the day before I picked up Annabel and we drove off into the wilderness. Lisa came with me to a bank downtown so I could sign the reams of pages and have them notarized.

A young, skinny clerk ushered us to a desk. He looked barely old enough to drink. "What have you got here?" he asked cheerfully.

"Oh, my divorce," I said, and then broke down crying. He looked from me to Lisa while she patted my hand and offered me tissues. She started crying too.

An hour later I cut up the shared credit card. Every time I pulled my new one out to fuel the big truck or buy more provisions, I practically hyperventilated.

By the time I picked up Annabel, I'd pulled myself together outwardly. But I was exhausted, and I knew at some point I'd have to tell her that her parents were divorced.

Now, steering our shared ducky, guiding us both into some unknown future, I felt grateful for the opportunity to spend time away from it all with her. I wanted to show her a side of me that was capable, brave, and forward-facing—literally, down a canyon—even if I didn't necessarily feel all those things. Sitting in the bright yellow boat and paddling slowly, I now felt prepared to understand that this trip might not be the fresh start I'd hoped for as much as a necessary reckoning with the past.

And there at the sandy launch site, waiting to send us off, had perched a great blue heron.

AT THIS POINT in his trip, Powell still had four boats. He was, as Stegner describes him, both egotistical and almost pathologically optimistic, "as single-minded as a buzz saw." Waiting for him, though, was Disaster Falls, one of the biggest and longest rapids in the river. At the top of the rapid, the boatman rowing a boat dubbed the No Name was unable to pull to shore in time to scout and portage the tricky bits. The boat was swept into the current like a wooden clog. It struck a boulder sideways and split in two. The men aboard survived the swim, as did a keg of whiskey, but precious food rations, guns, and clothing did not. At this point, George Bradley, a former army sergeant and fossil hunter who kept an acerbic secret journal of the voyage, scribbled, "If we succeed, it will be dumb luck, not good judgment that will do it."

By the time we got to Disaster, the water was flowing at a much lower volume than in Powell's time. We stopped to scout from shore and I felt the familiar butterflies in my stomach. The falls are actually a series of drops and rocks and tricky moves. Annabel joined a six-person raft and I switched to a single-person inflatable. Two teen boys took over one double inflatable, and another couple got in the second. One of the boy's moms mouthed "I love you" to him and then she smiled and mouthed it to me, too. We laughed, but we were all nervous. The head guide told us to follow his line. I used to hear this phrase a lot while kayaking with my husband through big rapids: *follow my line*. I also learned early on, the hard way, that it's not always a good idea. Since my childhood, I've known how to read water and I've gradually learned to paddle it the way that works for me. I rely less on brute power to move my boat and more on planning ahead, sometimes paddling backward or upstream to buy more time, and then keeping my boat straight and fast through the big water. Sometimes I look more like a manic wind-up toy than the studied, laconic-on-the-outside men I'm paddling with, and I'm okay with that.

As I watched the guide's raft entering Disaster, I thought he was too far right, where a large boat-eating hole was waiting but hard to see. I saw his raft disappear into it, get caught sideways for a perilous moment, and then kick out, minus the guide, who'd shot out like a surface-to-air missile. Following like ducklings, the boys flipped over fast. I made a swift recalculation and paddled hard to the left, where I had seen from shore a decent slot over a small pour-over. I made it and then looked back upstream: two overturned inflatables, five swimmers getting hauled into rafts, paddles everywhere. River carnage. I scooped up two paddles and ran Lower Disaster balancing them on my lap.

Edward Abbey, a desert rat who knew this country well, described big rapids as "vicious loveliness." (He also compared waves to heaving breasts, because that's the sort of writer he was.) The river was already reminding me of so many lessons I needed. Trust myself. Follow my own line. Move through fear. Turn trepidation into action.

After Disaster, the teen boys started following me through the rapids. Annabel was happy to jump in my boat. From the bow later that afternoon, she spotted two bighorn sheep and an osprey carrying a perfect fish in its talons as it flew overhead. She's always been the finder in the family, quiet, poised, on the watch.

On night two Annabel and I carried our gear high up to a flat sand ledge in a nook below rust-colored cliffs. Our girl cave sat in the basement of the world, among hardened layers of sediment nearly a billion years old. After a meal of salmon en papillote, we watched the brilliant stars through the narrow mail-slot of canyon walls above us. Cliff swallows dove overhead and we woke to the descending trill of canyon wrens, a song I've been listening to on these rivers since I was exactly my daughter's age.

A WRITER I ADMIRE, Caroline Paul, who has been a firefighter, luge runner, and paraglider pilot, describes in her books and essays how men and women think differently about fear. They both have it, but men tend to look at larger goals and benefits and believe what's fearful is worth pushing through. Society and their peers reward their curiosity, ambition, and risk-taking. But fear often holds women back. They're taught at an early age that someone is likely to outshine them, rescue them, soothe them, and comfort them if they give up or stay home.

The result: a bravery gap, one with lifelong implications. A 2014 survey of over 1,500 teenagers found that 92 percent believe bravery is important for meeting life goals. Yet only 50 percent of girls consider themselves brave, compared to 60 percent of boys. Worth noting is that the girls who identified as "explorers" had the highest bravery scores, were the most likely to participate in activities, and earned the highest grades in school.

Another survey of 2,000 women commissioned by the outdoor retailer REI found that 74 percent believed that the outdoors was a place where they felt free from the pressures of everyday life. Those who participated in adventure sports when they were girls were more likely to remain active as adults and were 20 percent more likely to place a high value on adventure later in life. (Juliette Gordon Low intuited this need. A widow who had been unhappily married to a philanderer, she founded the Girl Scouts in 1912.)

I didn't know of a survey looking at middle-aged women who participated in adventure sports as if their lives depended on it. But out here, I was starting to feel that my body was made of something other than thin, brittle paper. I was solidifying under a blanket of shoulder muscles, silt, birdsong, an unfailing solar system.

On night three, after a dinner of grilled steak and potatoes, I spoke

to Garth, a genial guide with a Cat Stevens beard, about my plans to continue down the river. No matter how hot it was, Garth wore a thick plaid
shirt. While the teens were playing Frisbee on a narrow beach, we pulled
out some river maps. In an earlier season, he had rowed through the lower
canyons. He told me about the steep takeout and where to find unmarked
panels of ancient rock art.

"Being alone is going to be scary," he said, looking up from the map.

"It is?" I asked, surprised to hear this coming from a brawny millen-
nial with Pleistocene facial hair. A *professional river runner*. "But it's all
flat water."

What did he know that I didn't?

ON THE MORNING of day four, we emerged from the curvaceous narrow cliffs of Whirlpool Canyon to paddle a long lazy stretch, one of many
that Powell named a "park" marking the wide-open country between
canyons. Then, after six miles of meandering river, we came upon Split
Mountain. A jumble of confused landscape, it rises from the river like a
mille-feuille cake, upside down, partly smashed, gashed down the middle.

One does not often see mountains sliced in half. One-armed Major
Powell was a geologist, but he was stumped by Split Mountain, or, rather,
by the river that ran straight through it instead of around it. Powell cor-
rectly concluded that the water channel came first, and the mountain
rose up slowly underneath it, leaving the persistent river in place as it cut
through. Later this remarkable uprising of ancient sea beds, which began
around 80 million years ago, would be called the Laramide orogeny. Saying it made my mouth feel full of river mud. Elsewhere, many of the layers
resemble, as our guide Zack put it, "an aging redhead," with bands of gray
resting atop deeper bands of red. But at Split Mountain, the order all goes
to hell.

Zack, who studies geology when he's not running rapids and grilling
steaks, gathered us up to talk about it. He was reverent. "This is a great
upheaval," he said, like a preacher in distressed times. "Things like this

don't just happen," he continued. "There had to be a big event, a fault. We are going from an anticline to a syncline!" He folded his map like a W, showing us the ground-level parks in the troughs, and the canyon walls and Uinta Mountains in the peaks.

Split Mountain: a big upheaval, faults. Something as solid as a land-form, as solid as a marriage, cleaved and smashed like a dropped cake. And now, a somewhat terrifying opportunity to examine it squarely at ground level. It was perfect. Proust wasn't a river runner, but he wrote that matters of the heart agitate and enrich us, and, in this way, produce "real geological upheavals of thought . . . twisted about one another, in giant and swollen groupings, Rage, Jealousy, Curiosity, Envy, Hate, Suf-fering, Pride, Astonishment, and Love."

I'd paddled the class III–IV gorge of Split Mountain several times before. The steepest and fastest section of the entire river, it always makes my stomach turn, mostly because I've run it in my kayak at high water, when the waves can be as big as a house. Now, in late July, the water was not so pushy. My daughter, gaining in confidence and hungry for adven-ture, wanted to pilot her own boat, following close behind me. Enter-ing Split, we adjusted our helmets and cinched our life jackets. "Send it!" called the boys, lining up behind us in their duckies.

We ran four ripping class-III rapids in a row, bobbing up and down like corks at sea, hearts pumping, and then it was over.

"That was so fun!" said Annabel. Then, with a bit of censure, "Mom! You took me so close to that hole!"

I pulled her boat alongside mine and we looked at each other through our sunglasses.

"Bel Bel," I said, leaning my arm over her bright yellow boat, "some-times you have to find your own line."

III.

As the Green continues its southward flow, it turns browner and milk-ier, picking up loads of sediment from crusty gray-and-tan sandstone.

The country grows even drier and lonelier, with few trees and few roads. By this point, day 43 for them, Major Powell and his men were growing bloody tired of the desert and its sparse offerings, especially since the bacon had long gone off and the flour had turned green. Powell wrote with Victorian flourish even when discouraged: "The walls are almost without vegetation. A few dwarf bushes are seen here and there, clinging to the rocks, and cedars grow from the crevices—not like the cedars of a land refreshed with rains, great cones bedecked with spray, but ugly clumps, like war clubs, beset with spines. We are minded to call this the Cañon of Desolation."

And so they did. Desolation Canyon, my next gig.

AFTER SPLIT MOUNTAIN, I'd left the river for one night to drive my daughter through the Uintah Basin and over the Tavaputs Plateau to the Salt Lake City airport for her flight home and to buy my share of food for this next section of river. The pickup already contained our raft and all the gear my brother Berkeley and I would need for the next eight days and 84 miles of river. He met me at the put-in, a forlorn dirt ramp called Sand Wash, hours from the nearest paved road, with beer and more ice. My sister, Violet, and her boyfriend, Kevin, who live in Los Angeles, flew into a desert airstrip in the middle of nowhere on a small chartered plane. Five other friends drove in from Montana and Colorado.

Back in January, I'd sent out a group email to friends to see who wanted to join me in entering the lottery for this section's permit. It's a big ask, not only because it's a long, strenuous trip in the wilderness, but because if you throw your name in and win, you're committed. To my delight, quite a few said yes. I'd like to think they showed up to support my half-baked post-divorce self-actualization—and there was an element of that—but mostly, these people just love running rivers.

At the ramp, Ranger Mick checked our permit and our gear, including life jackets, safety ropes, and a chemical toilet system for solid human waste. Because it was an unusually hot, dry summer, we wouldn't be allowed to build campfires. Already, smoke from California wildfires was clouding the western sky. The Bureau of Land Management and the Ute Indian Tribe of the Uintah and Ouray Reservation take their steward-ship roles seriously. For a $10 federal fee and a $50 tribal fee, our group would be able to paddle through one of the largest wildernesses in the continental US (a year after our trip, Congress would officially designate a narrow stretch of river corridor as "wilderness," protecting it from roads and drill rigs).

"I have two things to say to you," Mick told our gathered group. "One: don't pee on land, only in the river. Two: watch out for bears. Lock up all your food at night, even toothpaste."

We were a flotilla of three rubber rafts, two duckies, my little pack-raft, and one optional stand-up paddleboard that my friend Diane would gamely wrangle through surprisingly large waves. Of the nine of us out here, seven were divorced. The only two who weren't were Violet and Berkeley, who, still in their 30s, hadn't married. This demographic wasn't intentional; I had invited married couples too. Diane, my age, had been separated about as long as I had, and she brought her new boyfriend along. His raft sometimes deployed a large pink confection of a sun umbrella, so we took to calling him Cupcake. One of us, John, a policy wonk at a non-profit, joined at the last minute after his two-decade marriage splintered. He was taciturn the first few days. My friend Laura and I tried to (care-fully) psych him up.

"You should get out there and date!" said Laura loudly as we bounced through a rushing riffle in our small inflatables. "Join Bumble!"

"Bump Her?" he said, mishearing. "I don't want Bump Her."

"Maybe you do!" we yelled.

OUT IN THE desert in August, I was learning a new appreciation for monsoon season. My father had warned me. Over sandwiches in DC with him and my son before we headed out west, he said, "I'm not wor-ried about the rapids or the snakes. The biggest risk to you is going to be storms." We had run this section together decades before in open canoes at low water. I didn't remember big storms. I remembered jumping off rock ledges into the river, floating the current in our life jackets, hiking to petroglyphs. I remembered how special it felt to be my dad's bow person, part of a well-coordinated team. At 81, he was still watching out for me.

The first night in Desolation Canyon—or Deso, as river runners affectionately call it—began dark and clear. We set up tents but most of us slept outside them on the sandy shore under an incredible ribbon of stars. Then winds kicked up in the middle of the night, and rain started landing on our faces. I ran over to the kitchen to tuck away camp chairs and check the boat lines. I crawled into my tent and awoke a little later to what I thought was the roar of the river. *But we're not camping at a rapid*, I slowly remembered. I stuck my head out of the tent just in time to receive a blinding blast of sand to the face.

On day three, epic headwinds battered us not far from camp. The duckies could power through, but the big rafts were stuck in place despite the swift current. Diane's paddleboard launched like a rocket into the sky and came down near a raft, luckily, where someone caught it and tied it on. Finally we made it to camp. I had to pee, and ran downriver to squat in the shallow water. Halfway through, another gust came through and I watched my packraft tumble end over end up the shore. I instinctively rose up to chase it, bathing suit and shorts still around my ankles. My sister saved it. Lesson learned. *Always, always tie up your boat.* (Also: *pull up your pants.*)

NOT HAVING MY kids along on this leg created a different emotional dynamic. I was alone in my boat all day, sometimes within talking or water-gun-fighting distance of the other boats, but often drifting along in my own bubble, feeling a hint of what was to come as my solo neared. This was the place to prepare: the river section that has straddled every phase of my life, from my adolescence to my marriage to now, some kind of rebirth into an unexpected state in middle age that I was trying to understand.

Desolate: forsaken, damaged, lonely. In the Bible, the word refers to wastelands and also to women abandoned by their husbands. I wanted to travel boldly through this canyon and find something else. In *Les rites de passage*, published in 1909, the French ethnographer Arnold van Gennep described ritualistic solos—ceremonies across many cultures that typically mark passages from one status to another, such as adolescence or entering marriage. He wrote about three essential phases common to these rituals: separation/preparation, transition, and reincorporation. During the separation phase, one is removed—physically and ceremonially—from the former life. Sometimes heads are shorn, tattoos engraved, a sense of innocence relinquished. The transition or liminal phase often involves the journey itself, a trial of hardship and reflection, while the final phase incorporates a celebration of belonging and a welcoming to a new, wiser, but also more burdened state. You are a warrior now; you are a man, a woman, a wife. Why not an un-wife?

At first I'd resisted thinking of my river journey in the context of a "rite" or "quest," because those terms have been too easily appropriated by Castaneda-following groupies and seekers. I wasn't going to be smudging myself with sage or going hungry in hopes of seeing visions. But then I gave a talk at a high school in California and I met a teacher named Julie Barnes. Barnes runs Marin Academy's outdoor program, which includes a three-day wilderness solo for seniors. She told me how powerful the experience is for kids who are about to leave home for the first time. It strips away their usual distractions and supports, and "homes in on what they want to claim in their lives as they go through this transition. Who do they want to be as they step across this threshold and then step back into the world?" She said kids at this age are wired for deep challenges that match their energy. It's how their brains mature and learn. Without some sanctioned hardships, they risk becoming overly infantilized, delaying growing up and assuming responsibility for themselves and their communities, or they may take inappropriate, truly dangerous risks instead.

Women my age, too, are wired for transition. The writer and

ethnographer Howard Norman once told me about interviewing an Inuit woman who laid out her life story. He said to her, "I see everything but the years from when you were 50–54."

"Oh, I have no words for those years," she replied.

"Why not?"

"In those years," she told him, "I was a polar bear."

These are our polar bear years. Everything about us is changing, physically, hormonally, emotionally. Our shifting roles in culture, work, and family upend our identities, even without divorce. I told Barnes about my own reluctant transition, and my desire to mark it, recover from it, figure out what the hell comes next. She convinced me of the importance of setting clear intentions for my solo, and I found myself reviewing them during the quiet stretches on Deso.

What did I really want to do out here? I wanted, of course, to be fixed—to transform into a woman ready to take on the rest of her life, to launch my boat as a means of launching myself into a better future. But Barnes had warned me those expectations were naive and impatient. It was more realistic to solidify the goodbyes. I wanted—like Barnes's schoolkids—to individuate, in my case, away from my moribund, fossilized identity in a couple. To do that, I wanted to access my bravery. I wanted to transmute my fear into something else, something like what I had glimpsed paddling through Disaster Falls and Split Mountain: momentum, power, agency. I wanted to learn how to take care of myself and learn how to be alone. I wanted to cultivate beauty and experience awe. I wanted, finally, to say goodbye to my marriage.

Late in age, the poet Mary Oliver reflected on the three primary selves that she lived with: the childhood self, who is "with me in the present hour. It will be with me in the grave"; the social self, "fettered to a thousand notions of obligation"; and a third self, a sort of imaginative, generative awareness. "This self is out of love with the ordinary; it is out of love with time. It has a hunger for eternity."

A solo was a rare opportunity to let go of the obligated social self that had ruled my adult life. My social self, as I'd known it, was dissolving in any case. I was now interested in the other two. As I became less domesticated, perhaps it was natural to fall out of the love with the ordinary, to look to the wild.

EACH DAY, we traveled farther away from civilization and its material veneers and farther into Desolation Canyon. In places, the crumbly beige-and-red rock walls reach a mile into the sky, cutting deeper into the earth than even the Grand Canyon. The river looks and feels witchy wild. Fed by the Yampa, the last undammed river in the Colorado system, the Green is largely untamed. But I also know the ways it has been harassed, poked, throttled, and compromised. Climate change is drastically changing this system. The Colorado River Basin has lost 19 percent of its water in just the past 20 years, a change wrought by less precipitation and higher air temperatures. Squawfish, sometimes called white salmon, once migrated all the way from Wyoming to the Gulf of California on the Colorado and Green Rivers. Large, ungainly, and, like this river system, very old, they've been wiped out from the lower reaches, existing now only in minuscule remnant populations up here. Three other million-year-old fish species are also barely hanging on.

Since I first ran the river 35 years ago, oil and gas development has changed the land, the air quality, and the soundscape beyond the immediate river corridor. The canyon has lost its innocence, and with it, so have its close observers. Seeing the change in my lifetime, I felt more than one heartache.

"One of the penalties of an ecological education is that one lives alone in a world of wounds," wrote Aldo Leopold in his essay "The Round River." The area's Indigenous Peoples and their ancestors have been feeling the wounds of change for centuries, beginning with the loss of megafauna 13,000 years ago, the extreme droughts of 900 years ago, and the colonization, genocide, and land theft of the past 250 years, the legacies of which continue to this day. The antidote to the current ecological anguish is twofold: prevent the wounds as much as possible; and share the knowledge, and the grief.

A FEW DAYS from the end of the trip, the winds and rain ceased, but a new hazard emerged: intense heat. We didn't know how hot, but we speculated. Ninety-seven degrees? A hundred and three? One central irony of the desert is how desiccated a river can feel. On shore, the heat quivers off the pale soil and up the rocky walls. "What strange anomaly is a desert river?" marvels Ellen Meloy in *Raven's Exile: A Season on the Green River*. "In many stretches there are no banks, merely a precipitous jumble of wall and cliff, a boulder or sandbar to stand on but no place to park towns or pasture or croplands."

We were chugging through our water supplies. The ice in the coolers melted sooner than expected, rendering the gin-and-tonics lukewarm and the last dinner unexpectedly vegetarian. It was a pile of rice, beans, and salsa on tortillas, and it was delicious. After pulling over those last afternoons, we grew listless. We set up tarps for shade and collapsed under them on our sleeping pads, waiting for cooler air before setting up camp. But it didn't ever get cool. Diane and Cupcake would bring their camp chairs down into the river eddy and sit in them up to their waists in water.

With no chance of rain, we gave up setting up tents. One morning, rising from our sleep spots on the beach, someone noticed fresh bear tracks very close to where our heads had been. This bear, he or she, had walked back and forth several times. In Deso, many of the creatures hunt and forage at night. We were the only fools out in the high sun. My sister was unnerved, but I was not.

I liked feeling that we are not alone.

17

CONFLUENCE

Those who travel to mountaintops are half in love
with themselves and half in love with oblivion.
—ROBERT MACFARLANE,
MOUNTAINS OF THE MIND

We disembarked near the aptly named town of Green River, Utah, nearly a century after Buzz Holmstrom did the same. A depressive gas-station attendant and one of the first men to run this entire river alone, he'd pulled his wooden dory ashore here after many weeks on the water. Hoping to find a meal, and maybe a drink, he called it "the most miserable dilapidated one horse town I ever saw." Today, Green River is known for melons (grown with Green River irrigation), Mormons, and the John Wesley Powell River History Museum.

I hugged my brother and sister and friends goodbye in the wavering heat. It was, according to our freshly connected phones, 103 degrees.

Everyone had words of advice for my next leg of the trip. The strongest came from Dan, the most experienced of us. He put his hand on my shoulders and spoke two somber words: *Stay hydrated*.

I ducked into the museum, enjoying air-conditioning for the first time in two weeks. Amid replicas of Powell's boats—smaller and more delicate even than I expected—hung paraphernalia from other travelers. One was a large black-and-white photograph of Hollywood actress and folk musician Katie Lee, who first ran the Colorado through the Grand Canyon in 1953. She sang every night in exchange for a free trip. She returned many times with friends. In this photo, she stands fetchingly naked in a rock crevice in Glen Canyon, just upriver from the Grand, shortly before it was entombed by Lake Powell courtesy of the second-largest dam in the US. The photograph appears like a burst of estrogen emanating from the shrine to explorers. She was, she explained, "only doing what I always did in Glen Canyon . . . covering myself with mud, playing, singing, living with the Canyon." When the canyon was lost forever, she was disconsolate. Ed Abbey, who had also run this stretch, along with other rivers before they were flooded by dams, said later, "Every river I touch turns to heartbreak."

I drove the truck an hour south to Moab, Utah, where I'd leave it parked for the next two weeks while I paddled the Green's Labyrinth and Stillwater Canyons. I checked into a motel, showered, went out for tacos and the last cold beer I'd see for a long time. Back in my room, I made sure my emergency beacon was working and sent an email to some friends and family with the link to my satellite coordinates in case anyone wanted to follow my progress or make sure I wasn't a motionless signal, decomposing under a pile of driftwood. Then I paged through a field guide to local snakes and scorpions. Some were viciously poisonous. I wouldn't be able to phone or text anyone to ask what to do if I got stung. The emergency beacon offered only a last-ditch measure of protection. It was like the big red "help" button you give your co-workers for their desks, only it wasn't a joke. If pressed, a helicopter would supposedly appear, no questions asked. There was nothing subtle about it.

I called my kids. They were excited for me, and also a little concerned. There goes Mom, off in her . . . canoe. Then I called my friend Ann, with whom I've been running rivers for decades. She understood the desert perils. She also knew why I was here: to say goodbye to the relationship that had defined my life and to reclaim myself. She knew I felt alone. Soon we were both crying. I fell asleep that night with the book of scorpions splayed on the bed.

THE NEXT MORNING, I shopped for groceries, filled the cooler with ice and meat, topped off the water jugs, and packed the cooking gear, first-aid kit, and river bags. I ran to a store for extra matches, extra flip-flops, and that garish beach umbrella, which might be my only shade for much of the trip. It looked like a lot of stuff. I parked my truck at Tex's, an outfitter who'd be picking me up by jet boat precisely at 11 a.m. 13 days and 120 miles later, just before the river sweeps into Cataract Canyon and then the Grand. And then Craig the outfitter drove me back to Green River for my ill-fated evening launch.

That first night by the interstate after nearly tipping over was filled with dread and self-recrimination. What was I doing out here, alone, in the desert in August, with a freighter for a canoe? Even in gentle riffles, a canoe could capsize, and it would be very difficult if not impossible to swim this beast to shore by myself. Flipping was not an option. But now, acknowledging the possibility, I saw I would need to keep the beacon on my body constantly, along with some emergency food and tablets for purifying water. Being alone, I was realizing, was more than anything about not fucking up. I would have to tie the gear down at all times, secure the boat perfectly at night, camp away from washes where flash floods could appear with no warning, not cut myself, not break a limb, not light the beach on fire.

It was clear that while I might find moments of peace, I needed to be fully alert all the time. Hypervigilance is, I remembered, the physical state of loneliness. Now I was learning the literal, evolved reason for it: survival. We weren't supposed to be alone in the wilderness, and if we were, we needed every sense turned on high, every task-list made, followed, and double-checked. I'd been wanting to reinhabit myself by experiencing solitude, to turn my loneliness and grief into something more generative. But now I was worried that the opposite was more likely: turning solitude into loneliness. That night by the highway, I was feeling more vulnerable, not less.

I SLEPT POORLY and woke early, anxious to paddle many miles ahead of the mid-afternoon heat. The morning air was already stultifying. It felt good, though, to kneel in the boat and pull my paddle through the water. My friend the heron had welcomed me to camp the evening before, and I was delighted she was still flying around and landing alongside me throughout the morning. I have no idea if she was actually a she, but I took to calling her Gabbie, for GBH, great blue heron.

The river kept braiding. I tried to follow the bubbles in the current pointing to the channel with the highest water. Bubbles are the secret language of water. But sometimes there simply was no high water. Sometimes the canoe would run aground, and I'd have to get out and pull it by the bowline like walking a very fat, recalcitrant dog.

I still hoped to get rid of the heavy toilet. Late the second morning, I came upon a private boat ramp, where I was lucky to find a driver from Tex's dropping off some other canoeists. I hadn't seen anyone or heard another human voice since I'd launched 14 hours earlier.

"Are you Darren?" I asked a heavyset sunburned guy with a beard.

"I'm Darrell, his brother," he said. "Are you the solo lady?"

"Yep," I said. He told me the river was getting lower and there was some concern about the water taxi being able to get to my appointed pickup spot, but I shouldn't worry about that now. In the meantime, he agreed to take the behemoth toilet back to Moab for me. I had just enough special baggies called Wag Bags and a plastic drum with a lid for human waste, a still-legal but much lighter option.

"What do I do if I need to come out early?" I asked, measuring my options.

"That's very difficult," he said. "All the spots on the jet boat are reserved. But it helps if you cry a lot."

"I'm good at that!" I said, and waved goodbye.

WITH THE BOAT lighter, I felt better. I started to make good time in the mornings, resting sometimes to glide along in the slivers of shade against the shiny red cliffs, sip my green tea from my thermos, talk to Gabby, or some other Gabby, when she appeared. *Che cosa, Gabriella?* I was encountering plenty of awe: sunrises, the flitting dance of wrens, the striations of Triassic rock. The third morning a beaver swam around me and slapped his tail five times. The sounds cracked like gunshots. Was he issuing me a warning? Wanting a call-and-response? Beavers are ornery little characters. But I admired them for their resurgence. Once nearly wiped out from North American waters by trappers and then ranchers, they are finally now appreciated for their industriousness, utility, and charm. They were, I decided, worthy role models.

After making 12 or 15 miles by early afternoon, I'd scout for any tree. Most days, I got lucky and found one or two where I could string up my hammock. I'd swim in the now-lukewarm river and by 3 p.m. I'd become a bat, hanging lifeless. I drank a lot of water. I checked my blood sugar; it was good enough. I was finally learning to meditate. It was so quiet I could hear the inside of my head, literally, a subtle high-pitched whirring. Was this the normal sound of a human head? I'd never heard it before.

I found it a surprising relief not to talk to anyone. How much time we spend in social grooming! There was now no gossip, no small talk, no assuring, contriving, negotiating, coercing, managing, mollifying, texting, sexting, posting, replying, forwarding, liking. How much endless space there was in the day, suddenly. I started feeling less clenched by fear, though not unguarded. The human female's main predator is the human male. But with no roads here, I wasn't worried about that. In camp, I swam and lay about mostly unclothed, talking to myself, picking the silt out of my belly button. I liked it.

In a country that valorizes the lone hero and ascribes to self-reliance the highest virtues and achievements (looking at you, Emerson), the woman-alone-in-the-wilderness narrative is shockingly new. It is, perhaps rightly, the next frontier in female power. In most parts of the world, still, to be a woman traveling alone across remote country is to invite peril. Thanks to Cheryl Strayed's memoir *Wild*, there were now all sorts of women self-actualizing in nature. It has become almost as predictable a trope as the male version. But for my generation, it wasn't predictable at all. We grew up reading *The Right Stuff* about a bunch of dudes carousing, breaking the sound barrier, and then blasting off into space. To venture safely alone as a woman remains the rarest of privileges.

EDWARD ABBEY CALLED these canyons "one of the sweetest, brightest, grandest and loneliest of primitive regions still remaining in our America." Of course, he had the good sense to run this stretch in November. The fourth day finally brought some cloud cover and breezes. The temperature felt more like 93 than 103. Midday I stopped to climb a saddle in the cliffs to view the goosenecking river on both sides. There's a reason Major Powell named this canyon Labyrinth. He was both fascinated and exasperated. From a boat, it's disorienting. You paddle into the sun, then away from the sun, then to the side of the sun. "We go around a great bend to the right, five miles in length," Powell wrote, "and come back to a point within a quarter of a mile of where we started. Then we sweep around another great bend to the left, making a circuit of nine miles, and come back to a point within six hundred yards of the beginning of the bend." Ever to the point, his crewmate Sumner just scribbled, "River very crooked."

Mornings: drifting, paddling, tea, herons. Late mornings in the canoe, if it wasn't windy, I'd pop open the nine-dollar beach umbrella and sit beneath it like a duchess out for a carriage ride. I became fiercely attached to it. One breezy afternoon it blew out of its mooring, skittered across the river upside down like a spinning top, and then sank. I cried out. Then it popped right back up again as if it heard me. Thrilled, I managed to lean out across an eddy line and grab it without tipping over.

For the most part, the canoe, ever lighter (or was I just stronger?) moved pleasingly through the water. Of all the boats I paddle, a canoe is by far the most graceful and the one I have been paddling the longest. When I ran portions of the Green River as a teenager with my father, we shared a yellow canoe made and purchased in North Carolina. We named it Saga. Dad waxed on about it in a book he assembled for me about one of our trips: "Curious that this modern-day artifact is made with plastics of our industrial age, the wood of trees that started growing before the day of plastics, and the leather from animals whose very killing poses problems. It is a fine canoe and I love it dearly."

Paddling a canoe is like riding a bike. You move a limb, and the whole craft surges forward. This was the exact cause and effect I needed to feel, that I could be the pilot of my own craft. Maybe this is what Helen Fisher had meant all those months earlier in advising the heartbroken to create their own narratives, whatever they were. Maybe the story doesn't matter as much as the act of authoring it.

From above, a canoe is shaped like an almond, a pointed oval. The Italian word for that shape is *mandorla*, and it graces many cathedral doors and religious iconography, symbolizing the space between two states. In a Venn diagram of two circles, the section where they meet is the mandorla. It represents the tension of opposites, the space between the familiar and the unfamiliar, a threshold zone of both discomfort and growth. It's the dark space of a garden where you can't see what's sprouting yet; it's the sound of a rapid before you run it.

BUILDING ON VAN GENNEP'S rites-of-passage work, British anthropologist Victor Turner wrote about the liminal phase of ritual solos, indicating that they provide opportunity not just for personal transformation but for societal change because they upend social norms. You are no longer who you were and you don't yet fit into the world to come. Because participants exist temporarily outside of society, they invite new possibilities upon return. Wilderness amplifies this project, argues ecopsychologist Robert Greenway, because if you spend enough time in the wild (he believes a minimum of 10 days), it offers a "retreat from cultural dominance" and a dismantling of your normal construction of self and ego.

I was paddling my *mandorla* between two worlds. It demanded patience and a willingness to feel the edges, to turn over the rocks, to pay attention.

"Let yourself fall apart," my sister-in-law Lisa had said.

"Spend time out there thinking about what really went wrong," said my therapist.

"But isn't that ruminating?" I'd responded, already resisting it. "The same agonizing thoughts over and over?"

"It's not ruminating if you can find new insights," she pushed back.

"I don't trust my insights."

"That's what I'm here for," she said.

It was homework, and I needed to get on with it. I was dreaming about my ex-husband every night. He was waiting to be reckoned with.

NEAR THE MIDPOINT of my trip, I was aiming for a campsite my map said had trees. Because the tall cottonwoods are so rare in these canyons, they make camp easy to find. Paddling in, I noticed that across the river a giant slab of rock had flaked off the tall cliff and landed in the shape of a perfect heart, mottled and fractured around the edges. Bird poop dribbled over the top, but it was solid and intact, flashing like a neon sign over the water. I was delighted no one else had claimed the spot. There weren't many groups on the river. I'd seen a few, including some Boy Scouts who'd passed me earlier. I set up my hammock, meditated, and read C. S. Lewis on grief, written after his wife died. "No one ever told me that grief felt so like fear," he wrote. "I am not afraid, but the sensation is like being afraid. The same fluttering in the stomach, the same restlessness, the yawning."

If I couldn't find some measure of stillness here, I'd never find it. At this mottled-heart camp, I was beginning to inhabit the quiet. The heat and silence, the weight of ancient rock, the very slow river. I was stripped of disguises, clean as a bone. There was no place to hide from my memories.

Why was it so hard to confront my marriage? That evening, per a friend's suggestion, I wrote my ex a goodbye letter and thanked him for the many things I'd learned with him and the things we shared and the good ways we'd loved. I wrote about the ways I felt hurt and apologized for the things I'd done wrong. Sitting on a rock outcrop high above the river, I burned it in a frying pan: a marriage crisping like a slice of bacon. Crying, I cast the smoldering ash into the current. One blackened scrap flitted back to me. It contained just one word. "Sweet."

I found myself resisting saying goodbye to the marriage, to him, to the life I'd enjoyed. I opened a notebook. I made a list of all the things I loved about him, and then I wrote down all the things I didn't love about him. What I admired: his apparent ease, his sunniness, his competence, his enthusiasm for adventure. I loved our family unit, that these kind, tall, lanky humans were all my people, and I loved the happy delusion—such a delusion—that we were surrounded by a special force field of safety and fine weather. What I didn't admire: I won't enumerate his failings here except to say the biggest—the one I didn't miss at all—was his ability to make me feel that I wasn't quite worthy of him. That was bad enough, but the bigger problem was that I had started believing it. And as long as I still believed it, the worse my life looked alone.

It didn't help that my main work, per Julia, was supposed to be examining my role in the marriage's troubles. At camp, I probed deeply into my flaws, my failings, my resentments, and my neuroses. There were so many. I'd been impatient with his needs and incapable of effectively conveying my own. I was tired, tightly wound, always busy, not spontaneous enough. If he couldn't love me, maybe no else could either. My mood found reflection in the landscape, as it often does. What I saw, perched alone on a small ledge at the bottom of a cavernous Triassic bathtub, was a brittle emptiness.

I woke up the next morning feeling dizzy, with an upset stomach. Maybe my body was sending me a message about the necessity of purging the past, but it left me weak and achy. To be ill and alone is a new fright. I catastrophized. What if I get dehydrated? What if I get heat stroke? What if my brain suddenly swells and I'm too delirious to press the magic Eject button?

I swallowed some electrolytes and forged on to meet my resupply at the road that evening, where I figured I could hitch a ride out with Dave's grad students if I was in really bad shape. This was the one road in all this wilderness. By evening, I felt better, if still a little shaky. I set up a camp chair under a cottonwood to wait for Sara LoTemplio and Emily Scott, who soon arrived in a big university pickup bearing lettuce (yay, lettuce!), plenty of fresh water, and a chicken salad for dinner.

Company! We set up camp chairs and devoured the cool, tasty chicken. Both women were working on their doctorates in cognition and neural science in Dave Strayer's lab. In their 20s, they were keen to share their own outdoor adventures, as well as the work they were doing now, which often involved strapping portable electroencephalography machines to the heads of volunteers. Among the questions they were trying to answer was whether our brains behave differently in the wilderness as we attend to the elements of nature rather than to the directed demands of modern office life. For example, our task-focused frontal cortex appears to quiet down when we are outside while other parts of the brain, perhaps those associated with empathy and creativity, power up.

As we washed up and filled water bottles, they told me one of their current studies suggests that we may be processing external rewards differently after time outside. Volunteers who had spent a couple of days camping by a desert river produced weaker brain waves in a part of the anterior cingulate cortex linked to "reward positivity" while winning money in a gambling task. In other words, when immersed in nature compared to being inside, the volunteers didn't seem to care as much about winning. "One interpretation is they may be less motivated by external rewards in nature," said Emily. This interested me, because, as Johns Hopkins psychologist Sharon Kim had explained, it's helpful after adversity to identify less with labels like "divorced" and the ways that others judge you. Internal validation—the kind you create for yourself—is the kind that matters most for emotional health. As I was discovering outside, though, it was still plenty easy to generate internal criticism.

I was now halfway through the solo. Leaving the friendly faces and the dirt road behind the next morning, I paddled somberly for an hour and came upon a group of five preparing to leave their camp. As I got closer, I could see they all wore the same outfit—pale bodies and a dark patch in the middle. Then I realized they weren't wearing clothes at all. They appeared to be in their 60s and 70s. It was impossible not to laugh. They waved and I waved back. Another hour later, they passed me in their canoes. This time they wore shorts and T-shirts and we chatted across the water. Retired schoolteachers from Oregon, they had run this river many times. An hour after that, I paddled by them, sitting on the bank, eating lunch, naked again.

In mid-afternoon, they caught up and invited me to join them to hike to some ruins we were approaching on the left bank. "The route is a little confusing," said one man, who looked exactly like Jason Robards in *All the President's Men*. "We promise we'll be wearing clothes!" I joined them and, as we walked, chatted with one of the women.

"What brings you out here, alone?" she asked.

"Personal disaster," I said, and told her more.

"I wish I were alone!" she said, glancing back at Robards and rolling her eyes.

After our hot, hot walk, I joined them for a swim, naked of course, in the river.

———

IT WAS HARD to feel bleak frolicking with naked septuagenarians. But they were the last humans I would see for six days. Alone again, I doubted my fitness for love. And the doubts were deep. In cataloging my failings, I'd missed an important step, although I didn't know it at the time. I didn't yet realize it was okay to be broken, that it was even, perhaps, essential to becoming a more porous animal capable of far more real love than I had known was possible. It would still take some time for me to learn that our flaws are not the problem; rather, it is the failure to forgive them—in ourselves and in others—that trips up our hearts.

All I knew then was that I craved a good talk with my sister-in-law Lisa. What would she say to my sulking self? I could see her blond curls grow animated. She would tell me, sternly, that I was being an idiot. She would call me Ducks and point out that I still had plenty of good qualities, and then she would remind me, in a more jolly, frank way, that my ex was at least as big a loser as I was.

That's when I realized the real point of society. Without conversation, we simply have no check on our own dark thoughts. It is as simple as that. William James wrote, "The only enemy of any one of my truths is the rest of my truths." We need other people's truths even more, even if they're not true, if only because they are kinder.

We also need their goofing around. On Powell's grueling expeditions, the men sang and joked and made fun of one another's skivvies. *Conviviality*: such a happy word. After nearly two weeks without banter, my mood was flagging. I missed the registers of laughter. I missed human touch.

I came here to embrace being alone—or at least to face it—but the fact is I liked having other people around. I liked having a partner around. Relying at times on a kind someone—or on a close group of kind others—has always been our cellular superfuel.

––––––

THE WORDS *ADVENTURE* and *adversity* spring from different but similar roots. *Adventure* comes from Latin—and then French—meaning "about to arrive or happen." In Middle English, according to the *Oxford English Dictionary*, it meant "to commit to chance," to risk, to expose, to face a hazard. *Adversity* comes from the Latin, "to come to a twist or reversal." In a way, one is an inoculation against the other. Through adventure, we can learn to become comfortable with unforeseen outcomes. We practice adjusting our footing. We realize that, if we wait long enough, the weather changes.

One warm morning, a sinuous white-sand beach in the shade called out. I pulled over for some naked stretching and a swim. A raven became very voluble overhead. Its aria echoed through the rock walls upriver. *CAW caw caw caw CAW caw caw caw.* I lay facedown on a towel, mesmerized, heart to heart with the cool core of the canyon.

On this river I had laughed out loud often, watching the lizards scamper around and the beavers dart under the water, and waking one rainy morning to find the river had turned a brilliant and wholly unexpected shade of red in the night. I would miss these rhythms and surprises, but I found myself leaning toward home. I remembered one night on Lodore when I'd tried to sleep outside alone on the tarp while Annabel stayed in the tent. It grew windy. Sand was blowing everywhere. I'd felt agitated. Anxious about the upcoming solo and with my hair full of grit, I gathered up my stuff and trudged into the tent. I nestled in next to my daughter and slept hard.

I'd come here to find some time for hard reflection, and I'd done that. But I was ready for my dear ones.

MY LAST MORNING, I woke up early to paddle the last couple of miles to the Green's confluence with the Colorado, my pickup spot. With one water taxi showing up every few days and the water level dropping, this wasn't a day to sleep in. Today the river was swollen with new rain, so there would be no problem with the jet boat getting in. Marveling at the water's new deep-red shade, I was startled by how soon the confluence came up. I gasped. And then, naturally, cried. Two of the mightiest rivers in the West slowly twisted together like lovers from different directions: it felt like a power spot in the universe. I had seen it once before, from a mountain bike on an old canyon road up high on a butte, years ago with friends, in Canyonlands National Park. Now I was in the middle of it, baptized, washed through, birthed out.

By the time Powell and his men ended their trip another 300 miles south of here, only six of the original men and two of the boats remained. After nearly perishing in the rapids of the Grand Canyon, three men had thrown up their hands and decided to try walking out. Only one major rapid remained, but they didn't know that. The deserters never made it, murdered on the way out. Powell was quick to blame unknown members of the Shivwits tribe, but the killers were just as likely Mormon settlers suspicious that these unknown white men must be spies of the US government come to threaten their polygamous, antifederal empire. At the end of the 1,000-mile river journey, Sumner wrote, characteristically, "I find myself penniless and disgusted with the whole thing, sitting under a Mesquite bush in the sand. . . . I never want to see it again."

I paused for a long moment in the merging currents in what felt like the pulsing heart of the country. The rivers seemed particularly badly named—the Green River was running the color of raw salmon, while the Colorado, coming in from the east, flowed green. It was easy to feel the dynamism of the landscape, the sense that these rivers and their basin are always changing, whether over millions of years or the course of 24 hours. Life here is literally fluid. Another lesson from the river. Nothing stays the same for very long, and sometimes things change suddenly when we least expect it.

I paddled over to a wide sandy beach. A small group of canoeists was already waiting for the boat that would zoom us with surreal speed 48

miles up the Colorado to the outskirts of Moab and weak-beer civilization. They shouted greetings and waved me over. A sturdy woman with
gray hair and a Texas twang waded out to catch my bow. "What's a young
girl like you doing out here in this big country all by yourself?" she said,
walking my canoe to shore. If a man had asked that, I'd think I might have
trouble. But from her, it sounded wonderful, the start of a homecoming.

My lemon-yellow cowboy hat must have fooled her. She couldn't see
my legs were hairy in patches like a warthog.

"I'm 51!" I said.

"That sounds young to me!"

The beach lingered in morning shade. I took off the hat as I stepped
out of the boat. She reached out and pulled some twigs out of my hair. "I
haven't seen a mirror in 13 days," I said.

"Thirteen days!" she said, and then she hugged me. She pulled back,
inspected me, and then hugged me again.

18

THE HAPPINESS THAT MATTERS
Social Well-Being

At the heart of solitude lies a paradox:
look long and hard enough at yourself
in isolation and suddenly you will see
the rest of humanity staring back.
—STEPHEN BATCHELOR, *THE ART OF SOLITUDE*

I picked up my scalding pickup, threw out the soggy leftover salami, and turned my thoughts to shaving as soon as possible. I called my family. I told my brother-in-law, Peter, a Buddhist who is well practiced in silent retreats, about my progress in learning meditation, but also about my grim days cataloging my interpersonal flaws.

"Nooooo, Florence," he said. "On a solo with no social support, you

are not supposed to work through really difficult thoughts. Those are supposed to happen when you get home."

Oh.

Lisa censured me too. "Oh, Ducks," she said. "You internalized that
macho expedition crap. You should have just gone biking with me on
the prairie."

Maybe she was right. I had to concede, and it was hard, that maybe
I'd overreached with my expectations for healing on the solo trip. I'd put
so much faith in the redemptive powers of nature that I'd neglected the
biggest lesson I learned on the backpacking trip with the trafficking survivors. Nature can help you prepare for healing, but it doesn't do all its
work automatically. Especially when there's emotional trauma involved,
nature's best offering might be the peaceful kind: helping you calm your
nervous system enough to flex your imagination. But that peace is best
found in the context of safety, in a setting with supportive people. I'd
bought into the idea that I could tough out my problems by (literally)
toughing out my trip. It's a common mistake, founded in early military-
style wilderness programs and even the Boy Scouts, according to Denise
Mitten, the adventure professor from Prescott College.

"People always talk about the combination of adventure, risk, and
challenge, but I don't think that's the agent for change," she told me.
"There's still a patriarchal creep in a lot of these groups of getting out of
your comfort zone, but in general they are geared toward people who are
already comfortable. What about people who aren't comfortable? Nature
should put you in a comfort zone."

Being alone in the wild—while it had its glorious and liberating
moments—did not turn out to be so comfortable. And so it did not fully
yank me out of the dark recesses of my own head.

I thought about Emerson, and America's favorite river runner, Huck
Finn, and how the mythology of self-reliance seems like a celebration of
being free from social rules. Too often, though, this pursuit estranges us

from each other. Ultimately, Huck's story isn't about a boy and a river; it's about his friendship with Jim. Our lives and our systems and our communities are deeply interconnected, and to flee from that is to flee from our responsibility to care.

I TRIED TO explain all this to Steve Cole. About a week after I returned to civilization to hug my kids and ready them for the new school year, I flew back west to Los Angeles to proffer Cole my next vials of blood for analysis. I told him I didn't exactly feel cured. I'd thought about a lot, I'd processed pain, I'd cried, I'd grown physically stronger, I'd impressed my teenagers and myself. Ironically, all the motion had helped me—finally—settle into feeling okay with a measure of stillness. These were big wins. Had I accessed a little more bravery? Maybe. But had I burned off my stress? Had I come up with a new narrative and a neat sense of closure? Was I now mentally fortified to be possibly alone forever? Only partly.

Cole nodded, considering. We were eating lunch at a sandwich shop near campus. He ordered a bicolor soup served in a wide bowl. Peas and carrots swirled together in the exact pattern of yin and yang. He dug in. Cole is, by his own admission, not a nature guy. He's a people guy who has made the study of the bodily markers of loneliness his life's work. But as a psychologist, he appreciates the body of research on the benefits of nature. "We are fundamentally creatures of the dirt," he said. "It's not a bad way to reground yourself." It might not, however, be a great way to reboot your immune system if you're alone, he continued, waving his spoon. I told him about the fear, the vigilance, the constant need to check and cross-check my survival systems.

Would this stress be evident in my blood work? He reminded me that stress is not the same thing as misery. In fact, we need a measure of stress to get important things done and even just to get out of bed in the morning. *Happiness*, the way we typically think about it—as something easy and pleasurable—fails to capture what really fulfills us. It remained to be seen whether my cells would record any fulfillment or voice opprobrium the way Peter and Lisa had.

Cole has been fascinated by the way the field of psychology categorizes happiness ever since he collaborated on a study, published in 2013, with Barbara Fredrickson from the University of North Carolina at Chapel Hill. Typically, psychologists tout positive emotions, things like joy, inspiration, amusement, gratitude. These emotions make us feel better. They put a spring in our step. Promoting social bonds, they fall into the category of "hedonic" happiness, the kind derived from experiencing pleasure and avoiding pain. The Greek philosopher Epicurus was a big proponent of this one. Aristotle, on the other hand, preferred eudaimonic happiness. Impossible to both spell and pronounce, it represents a fairly straightforward idea of a longer game: fulfillment derives not from feeling good but from striving for purpose, and the idea that life in general, and your life in particular, means something, even in the face of adversity. To be fair, Epicurus believed both were possible at the same time. Why not strive for virtue while nipping into a nice pot of cheese?

Hedonic versus eudaimonic. Contrary kinds of happiness, like the shapes in Cole's soup. For the 2013 study, Cole put them head-to-head. The Fredrickson team gave questionnaires to 80 healthy volunteers to measure how much of each kind of happiness they reported. Cole measured their blood for his favorite suite of 53 inflammation and type I interferon–related genes, a package he calls CTRA, for conserved transcriptional response to adversity. He found an unexpected difference: although people with higher hedonic happiness appeared to be "happy," they displayed a worse immune profile—the kind that could put them at higher risk for getting sick—than people high in eudaimonic happiness. The Aristotelians might be exhausted and surly and lacking cucumber-infused face cream, but their purpose-driven bodies were more likely to hold up in the end.

Cole went on to collaborate with researchers on numerous other studies, including one examining male workers in a large Japanese corporation. The Japanese work famously long hours and lead stressful lives. Yet the ones with the highest eudaimonic well-being—those who felt most enmeshed in and fulfilled by the collective office enterprise—showed a 42 percent better gene-expression profile than those with lower levels of that type of well-being.

"Nothing we have looked at on the positive side of life can hold a candle to eudaimonic happiness in terms of changing gene expression," Cole said. "You might not always be having a fun, mirthful experience of life, but when we look at physiological data, you could be doing well." And forget Groupons for facials. What is today marketed as "self-care" is, he adds, "an unproductive response."

In another study, Cole and colleagues sampled blood from 108 retired adults in the US and assessed the participants for social isolation, loneliness, and eudaimonia. Supporting earlier studies, loneliness tended to worsen the CTRA genes while eudaimonia improved them. What was surprising, though, was that in individuals who reported being high in both loneliness and eudaimonia, the gene expression still looked pretty great. In those individuals, having meaning and purpose outflanked even the scourge of loneliness.

"Social well-being," said Cole, referring to a measure of eudaimonia, "is the strongest correlate of favorable biology in the positive psychology spectrum." This is true even in the hedonistic US. He pointed out that Americans can be plenty eudaimonic. "They are ridiculously hardworking people and not just because of a puritan work ethic," he said, "but because the American dream still lives."

Cole believes this research holds the key to people's health in an imperfect world. "The best antidote to loneliness is mission, not togetherness," he said. Simply being with other people is rarely enough to make people feel fulfilled. And many missions tend to bring people together anyway, compounding the benefits. I told Cole I struggled with terms like *purpose-driven life* because they're often paired with religious overtones and implied judgment. Cole is aware of this response, and he laments it.

"Western society has lost faith in morality and ethics in favor of the pursuit of individual health and well-being," he said. After all, health should be a means to an end, which is, essentially, to make some meaningful stuff happen, not just for you but for others. Currently, Cole believes, we're experiencing a clash between health and purpose. We privilege individual goals over collective goals, and we're the worse for it. "We've lost a sense of why we should do the things we do," he said.

The most successful interventions for loneliness, he continued, appear to be the ones that combine volunteering with community. He told me about one program for older women his lab recently studied. Generation Xchange is a partnership between the Division of Geriatrics in the David Geffen School of Medicine at UCLA and the Los Angeles Unified School District. For the study, it placed two dozen healthy volunteers (average age 68, predominantly Black, half of them single) in third-grade classrooms in South Central LA. The women offered math and reading support to the children 10 hours a week for nine months, during which they supplied questionnaires and blood samples to researchers. Not all the women rated themselves as happier by the end of the year, but their social well-being scores improved significantly. So did their healthful gene expression. And the magnitude of the change in blood work corresponded closely to their increases in eudaimonic (but not hedonic) happiness.

Psychologists track eudaimonia in two categories: the first is psychological well-being, which includes self-acceptance, personal growth, purpose in life, positive relations with others, autonomy, and mastery over external environment. The second is social well-being. Largely based on the work of Émile Durkheim, it captures our perceptions of our public and social lives, including factors like social integration, social contribution, social optimism, and social acceptance.

When Cole gave me his favorite questionnaire to measure these attributes, I scored high on the first category, psychological well-being. But I didn't score very high on social well-being, because I couldn't wholeheartedly agree with the statement that "our society is a good place, or is becoming a better place, for all people." This was several long years into the Trump administration; racial and social injustices were increasingly evident; America felt more politically and economically divided then ever; 40 years of environmental protections were unraveling and carbon was rising unfettered into the atmosphere. I didn't even know a global pandemic lay just around the corner. "It is hard to sustain general social optimism these days," Cole conceded. Yet a strong belief in social well-being, he repeated, is the single best predictor of favorable gene expression. He recommended I give it a try.

Global citizenship, he was implying, is like a successful long-term relationship; perhaps some flaws are best overlooked. If we can muster some optimism about the state of the world, and even better, if we feel we are actively contributing to its improvement, our genes will thank us. He was not suggesting that we merely paper over our disappointments and anger at society's failings, but that we harness them.

STEVE COLE CALLED me some months later with the before-and-after-wilderness blood-test results. As a reminder, the first sample Cole analyzed (let's call it Time One) was taken about nine months after splitsville, and the second sample (Time Two) about five months later, just after the river trip. It had been just over a year from the end of the marriage.

Cole began by saying the analysis was a bit complicated. He'd examined two particular types of genes: those related to inflammation, and those related to viral defense. Quick refresh: lonely people tend to express more inflammation as their bodies prepare for fleeing and fighting and getting injured alone in the jungle, and they express a reduced ability to fight viruses, which are spread in groups. After analyzing my blood for his favorite suite of inflammation genes (or more accurately, the transcriptional factors controlling those genes), he said my immune profile really did not look much better after the river trip than before it.

I was disappointed, but not totally surprised.

"Don't be too worried," Cole reassured me, "it's not like you look like you're about to keel over." But I did, he said, look like someone going through a hard time.

The problem seemed to lie in the antiviral genes, explained Cole. Based on my transcription factors, he saw that my stress neurotransmitters were the likely culprits inhibiting better progress. Noradrenaline gets released by the sympathetic nerve fibers when we are under threat. It then spills out into the bloodstream. Cole's lab can see the activity of genes sensitive to that signal.

"In your blood," he said, "there was not a ton of inflammation per se, but there was still a big signature of stress. These changes in your life are definitely filtering down to the molecular level in your body."

"Do my cells still look like those of a lonely person?" I asked, wincing a bit.

"Yeah," he said. "I would say so."

TRUTH SERUM, PART TWO

Maybe the only gift is a chance to inquire,
to know nothing for certain. An inheritance
of wonder and nothing more.
—WILLIAM LEAST HEAT-MOON, *BLUE HIGHWAYS*

As the weeks and months went by after the river trip, I considered more what had happened to me out there in the wilderness and why it hadn't dented my transcription factors. Yes, I gained a stronger sense of my own mettle. I could in fact paddle my own boat just fine. But I also realized that I didn't want to steer alone forever. I had dutifully flambéd that goodbye letter to my ex, but I was, many months later, still grieving my losses. The river miles had helped float me farther away from the shattering pain of heartbreak, but the destination remained stubbornly elusive. Wherever I was going, I certainly hadn't arrived. I still remained fearful of what would come next. Where was the story I was supposed to be creating, and how would the heroine behave? I wasn't sure. My trip hadn't delivered resolution.

Maybe I had been asking myself the wrong questions. Instead of "Why did I get dumped?" and "What do I do now?" I should have been asking, "What am I learning, and what do I want?" And, per Cole's insights, how can I best help others? There were opportunities out there, not just threats.

I thought of awe researcher Michelle Shiota, who told me that awe-states open a window for learning and for shifting your self-concept. She also told me that people who are very awe-prone exhibit a low need for what she called "cognitive closure." People high in the need for cognitive closure, she explained, prefer easy categories; they want to find clear reasons that something happens and they want to know what it means. They tend to prefer someone to blame and someone to follow. They are more likely to believe conspiracy theories and to feel comfortable with highly ordered and hierarchical systems, with rules, labels, and chains of command. People who enjoy experiencing awe, on the other hand, seem to accept that they don't understand everything. They don't mind some ambiguity. For them, the world is full of beautiful mysteries, or at least interesting mysteries.

Curious to see where I fit in the spectrum of need for cognitive closure, I filled out the 47-item questionnaire Shiota uses. It is called, helpfully, the Need for Cognitive Closure scale. Items include statements like "When dining out, I like to go to places where I have been before so that I know what to expect," "When faced with a problem I usually see the one best solution very quickly," "I prefer to socialize with familiar friends because I know what to expect from them," and "I like to know what people are thinking all the time." You rate each statement on a scale from "strongly disagree" to "strongly agree." My need for closure, according to the scores: middling.

I was beginning to confront the possibility that after heartbreak there is no big arrival signpost. Instead of trying so hard to find closure, maybe I needed to work more on becoming a person who didn't need it quite so badly.

I'D BEEN PADDLING canyons all my life. Perhaps all this awe I'd been getting was too . . . familiar. I started wondering about something a little more technicolor. Traveling to Northern California for some podcast work, I visited the man most responsible—after philosopher Edmund Burke—for putting awe on the map as an emotion worthy of our attention. (In 1756, Burke, who loved waterfalls and thunderstorms, delineated the sublime as a powerful phenomenon characterized by both amazement and mystery, eliciting wonder, fear, and even terror.)

Dacher Keltner of the University of California, Berkeley—along with colleague Jonathan Haidt, now at New York University—wrote an influential theoretical paper in 2003 that attempted to define awe and to bring it to the attention of other social psychologists. Awe, they explained, is a quintessential human emotion that "can change the course of a life in profound and permanent ways," they wrote. "Yet the field of emotion research is almost silent with respect to awe."

The son of a Buddhist artist and an English professor specializing in Romanticism, Keltner spent part of his youth wondering what to do with himself. His parents, he said, convinced him that "our best attempts at the good life are found in bursts of passion." He found himself most moved at punk concerts, which he started attending in high school. He loved the mosh pit, the sea of humanity thumping in unison, the presence of musicians like Iggy Pop. "He's full of originality and truth," Keltner told me. But studying psychology in graduate school, he was uncertain of his future, not publishing much, and feeling lost. So he wrote to Iggy. "And he wrote me back!" he said, laughing. "In the letter, he said, 'Good luck getting access to the young skulls at the University of Wisconsin.' The next sentence was, 'I dig teachers.'"

Keltner lifted his sweater to reveal an Iggy Pop T-shirt. He has seven of them. In his mid-50s, Keltner has the longish blond hair of a surfer and the laid-back demeanor of one, too. He's also a driven academic and the founder of the university's Greater Good Science Center, which is dedicated to investigating and publicizing methods for improving people's mental health. He rearranged his clothing. I wondered if all psych professors are secretly a little unhinged.

"So, yeah, after that I'm like, I'm going to be a teacher," he continued. "This is it. You know, it's important, when we talk about awe, to remember that there are these very idiosyncratic personal sources that are essential."

By now I was feeling pretty comfortable with Keltner, whose work I'd admired for several years.

"Can I ask you a question?" I asked. He nodded. I told him about my divorce. Then I asked him—because, after all, this was Berkeley and yes, because of that hair—what he thought about psychedelic mushrooms as a gateway to awe. I'd heard that among other things, like melting your face in a mirror, they could induce feelings of awe. I'd never tried hallucinogens before, and I thought Keltner might have something interesting to say.

"Do you think taking mushrooms might help me with heartbreak?" I asked.

He mused, and he smiled. "I was a really high-strung young person," he said. "I was going through the heartbreak of my family divorcing. And then the next year I went to college at UC Santa Barbara and I did mushrooms and LSD. My experiences changed my life and gave me access to the wisdom of awe, which is that there are big patterns out there that you're a part of. And the science, we've already alluded to it or described it. It can help you heal trauma. It makes you less stressed. It gives you perspective on things. And it is clearly a very deep source of change."

"So how do hallucinogens help with heartbreak?" I asked.

"There's this thinking in the psychological literature that heartbreak and rejection, whether it's seventh graders getting bullied or you've broken up with your partner, what happens is the self feels jeopardized, under assault, and threatened. And that's the core mechanism, right? And in fact that core mechanism of the self is vulnerable." I thought of the nor-adrenaline still flooding my cells as he continued. "It's part depression, and part certain kinds of anxiety and shame. And awe rearranges that. It's like, hey, the self's a part of a lot of things. And it just gives you these contextualizations that really help."

A graduate student knocked on the door. My time was up. "We can't forget the upside to awe," he continued, "that there are big, beautiful things out there that you are part of." He looked at me. "It has to be in the right context with the right people, you know? Okay," he said, "yeah. I would do it."

20

THE DIVORCE DRUG

To whom it may concern: It is springtime.
—KURT VONNEGUT, *SLAPSTICK,*
OR, LONESOME NO MORE!

I'd been reading accounts of life-changing psychedelic experiences and of unreasoned government fearmongering. Microdosing and megadosing a variety of substances were helping some people with chronic depression, grief, addiction, and fear of dying. They were credited with saving marriages, healing psychic wounds, renewing life's purpose.

In general, I'm skeptical of anything that's supposed to cure everything. A friend of mine who has suffered many years from depression tried a round of expensive sessions with ketamine, a psychedelic compound recently approved by the FDA for use in clinical settings. The internet is full of reports, including by journalists, of transformative experiences.

"It made me feel sunnier for a couple of weeks, but that was it," said my friend. And there are still many risks associated with these substances. All drugs, even good ones, have unwanted and unpleasant side effects. Bad ones for MDMA—otherwise known as Ecstasy, or Molly—include increased blood pressure and heart rate, temperature dysregulation, nausea, severe dehydration, diminished cognition with long-term use, acute kidney failure, and death. And that's with a pure dose not adulterated by the poisons sometimes found in street versions.

I'd heard that MDMA was being used (still mostly underground) in therapeutic settings during trauma counseling and also in couples counseling. Then I heard it was starting to be used in divorce counseling as well. An uncoupling drug? Now I was really interested, but I had questions. This seemed counterintuitive. Known as an "empathogen," MDMA is supposed to "open you up" and fill you with love, which you may not be keen to feel for someone who just cut off your health insurance. But it is also supposed to make painful events more tolerable, and to seed you with compassion for the circumstances and even the people who have hurt you. Perhaps this is because the drug is known to cause a cranial release of serotonin (associated with well-being), as well as the love-cuddle-breastfeeding neurohormones oxytocin and prolactin. The idea is that with a skilled therapist, MDMA could perhaps help with acceptance. It could help you move on.

My friend Blanche gave me a glimpse of its possibilities. A therapist herself in Portland, Oregon, she'd been living with her boyfriend for years, through much of her 30s and early 40s. He kept reassuring her he wanted to have children, but then he would find excuses to put it off. Recently, he told her he just couldn't do it and ended the relationship. She wanted a baby as much as she wanted him, and now, at 44, she had neither. At first she wondered why she would want to open her heart to him in therapy. But her MDMA-assisted session, led by former Zen therapists, was, she said, powerful. She and her ex went together, for a session that lasted several hours and included two half doses of MDMA.

"I had this idea that opening my heart would somehow make me open to being hurt, but it was really more of an openness to the truth," she explained. In the session, "all the ways we protect ourselves dropped off. I really could see the ways in which we weren't supposed to be together anymore, and it made it easier to come to terms with it without having to hate him to assuage my own hurt. It has honestly maintained over months."

A word like *truth* makes me hesitant, but I also understand it sits at the core of other therapeutic traditions that I am more comfortable with, like talk therapy. Even when "truth" can feel elusive and capricious, it's a helpful construct in a given moment until another truth becomes available. Blanche told me about a therapist she knows in Portland who works with not only MDMA but also psilocybin—magic mushrooms— *at the same time*. Maybe I could both assuage my hurt and have a bang-up awe experience.

I gave Judith a call, and I immediately liked her. I had already interviewed several people who seemed flakier and were more expensive. She emphasized she was not a guide; she was a therapist. I would not be able to use her real name. She might or might not feel comfortable enough to work with me, as she was new to this particular focus of her practice and wanted to make sure we were a good fit. I liked her smart and professional mien.

She explained how it might work. "The thing about MDMA is that it's much more like a therapy session, a super sweet and intense therapy session," she said, in which you contact your "heart and true self and see things that are blocking you." The MDMA acts like an insurance policy, she said, keeping you from going to a really dark place once the mushrooms kick in.

I signed up. In the intervening weeks before flying to Portland, I avoided reading too much about other people's psychedelic experiences (or even reading the studies). I didn't want my subconscious to borrow other people's images. A while earlier I had, however, read Michael Pollan's description of his psilocybin trip. In *How to Change Your Mind*, a look at the history and therapeutic potential of psychedelics, he writes of the world and himself disintegrating (in a pleasant way) into Post-it Notes. That's what happens when a journalist trips.

MY LYFT DRIVER from the airport was, rather perfectly, playing the Grateful Dead. Portland is like Boulder, but bigger and scruffier and way less manicured. People in Portland seem proud of the rough edges, the overgrown lots, and the rain that makes their yards grow huge and unwieldy. They exercise less and scowl more. They're not afraid to look morose, pudgy, resentful, to live in school buses. I liked it.

Judith lives in a graceful old bungalow several miles from downtown, filled with houseplants, Persian rugs, and an overstuffed couch. In her late 50s, Judith herself is tall and lean, her look part librarian, part yoga instructor. Wearing jeans and a T-shirt, she welcomed me in the morning with tea. We'd already done an "intake" session by phone, in which she wanted to make sure I wasn't taking antidepressants, and that I had a support network including a therapist at home who could help me process whatever kind of experience I had. We discussed medical issues and dietary restrictions. We talked about my intentions and hopes. She asked me to bring photographs of my ex and whoever else I wanted to conjure during my experience. I had three: my ex, my mother, and a man I'd starting dating a few weeks after my river trip. He was kind and attentive, but I had some questions about how I felt about the relationship. Would my subconscious mind yield any fresh insights?

At about 10 a.m., she mixed 120 grams of translucent MDMA crystals into a glass of lemon water and gave me what she said were a couple of neuroprotective supplements. I happily chugged down the water, expecting to soon feel blissed out and open. Within 10 or 15 minutes, I started to feel the room moving. "Am I supposed to be feeling spacey and woozy?" I asked.

She shrugged. "Everyone feels different!"

We went outside to her garden, and I told her a little more about the context of the divorce. But my face starting flushing very hot. My cheeks and jaw were aflame. My heart beat very, very fast and my stomach fell somewhere far below where it was supposed to be. I wasn't sure if I was now having a panic attack or if this was the drug doing what it does. I described it to her and said I would like to put an ice pack on my head.

"Okay, let's go inside," she said, helping me up some stairs. "I guess I should have given you two doses further apart rather than one big one." It was, I would learn later, a relatively large dose to take all at once. Too late now, I thought. It felt scary, like entering the current of a class-V rapid where there is no way out except through. I was committed now. And freaking out. Why did I do this? I hate losing control. I hate feeling like I'm going to throw up. I hate not feeling well. I vaguely recalled that MDMA could cause cardio-related side effects. In fact, journalist Michael Pollan's cardiologist told him not to take this drug. Was I going to have a heart attack? Had I made a big mistake?

She guided me onto a twin futon mattress topped by a sheet on the floor of her office. She went to the kitchen to fetch a bag of frozen peas and a bag of frozen corn. I lay down and stared at the paprika-colored walls, trying to breathe. When she returned, I placed one bag on my forehead and another on my stomach. I wished she could have brought me a full frozen side of beef. This was the hot flash of all hot flashes.

"Is it okay if I hold your hand?" she asked, kneeling beside me.

"Yes, good," I said. "Can we do some breathing?"

She told me to exhale slowly and hold my breath for a moment, and then breathe in and repeat. It felt good, and soon my pulse slowed. Weirdly, my cheeks were still burning, but now my fingers and toes felt like they were shooting out cold streams of water. This was not what I was expecting. And I hadn't even taken the mushrooms yet. I was thinking maybe I wouldn't, this was all I could handle.

She said she could lie down next to me if I liked, and I nodded and scooted over. She told me I was going to be okay. I grasped her bent knee through her jeans and she held my arm with her warm, smooth hand. I slowly felt more peaceful. At some point—I had little sense of time—I interlaced my fingers with Judith's and placed them on my bare stomach.

"I feel like a tree," I said. The image was very clear to me. My ex, I explained, was a vine. He was both embedded in my trunk, part of the cambium, but also constricting it. He was basically a strangler fig. After 32 years, he was a core part of my identity, and there were a lot of good parts as we grew together. We'd parented these beautiful children. He'd taught me many outdoor skills. We'd become a part of each other's loving families. "But now," I told her, "I need to keep growing on my own, into the canopy and into the light." I pictured large green maple leaves terracing outward and upward. It was lovely. Judith started taking notes.

"Can I untwine from him and still keep myself?" I asked.

I didn't wait for an answer. "He didn't love my dark side," I said.

But it was clear to me that my bumps and my darkness were part of my tree-ness. I valued them. "I didn't get to express and explore those parts enough," I chattered on. "I want to grow up around him and beyond him." My body felt lighter on the futon. "I have so much gratitude for the opportunity to become more myself," I added. It was the MDMA talking, certainly, but these insights seemed, in their corny yet undeniable way, very beautiful.

"How does this make you feel?" she asked. I thought about that and breathed.

"Unclutching," I said.

WE HAD A decision to make about the psilocybin, which was infused in a glass of purple tea beside me. It was best to take it while still riding the open glory wave of the MDMA, Judith said. I could take none, or half, or all.

I decided on half, and drank it. We cued up a playlist I'd given her and I put on an eye mask. Judith left the room to get herself a snack. I was drawn to the mason jar holding the remainder of the tea. Pungent and precious, it beckoned to me. I pulled it to me and drank it all down.

I giggled under the eye mask. "I took it all!" I told her when I heard her return.

The music took over my world. I could see sound waves separating and coming together in sequences of coherent color and light. At one point individual notes melted and dripped onto my face. A friend had put together my playlist. I told him I wanted the songs to be beautiful and mysterious, not dark and not simple. He included some Debussy and Beethoven (Piano Sonata no. 14 in C-Sharp), but mostly the ambient instrumentals of Brian Eno, a founder of Roxy Music and producer of U2, the Talking Heads, and David Bowie. The tracks, with names like "Among Fields of Crystal" and "The Chill Air," seemed uniquely engineered for tripping, with their synthetic piano, environmental elements, long atonal reverbs, and luminous floating notes. It's not for everyone. *Q* magazine once wrote, "Eno makes a muted synthesizer go bong for a couple of minutes short of an hour, which leaves you asking, is that the new Brian Eno album or is the fridge playing up again?"

I was part of the music; the music was part of me. I was flying through space, a filament of light, among thousands of other filaments of light. All of humanity and the whole world were indistinguishable colorful filaments, moving in murmurations. I couldn't tell which of the filaments was me, and it didn't matter. Sometimes two filaments would align perfectly and swim in parallel for a bit. It was clear to me these were our human relationships, sexual and otherwise. Our frequencies would line up and vibrate together, and then we would swim off and maybe find another pairing for a while. (I now see why the psychedelic sixties spawned free love.) There was no sadness, no grasping, no regret, not even something I would identify as love. Those were all just emotions, and they didn't really register. What was left were beauty and motion and light. I also saw the dome of the world as an intricately woven basket. You couldn't tell where one strand started and the next began.

Then I was deep inside a vertical curtain of beads, and the beads hung as strands of marble-like molecules interspersed by empty black space.

These molecules were all life and matter. There was no judgment, just physics. We were all just molecules. Our emotions, too: just molecules. This became amusing. How funny that we take ourselves and our feelings so seriously. *People, we are all just molecules.* My ego had fully disintegrated, just like Pollan's book suggested it would, and it was very pleasant. But the self occasionally returned as a voice of commentary and a point of view floating through the universe. At one point Donald Trump's head rolled in, and I said, *What are you doing in my trip? You are not invited*, and I drop-kicked him out. I came upon some demonic ghostlike creatures in dark corners, and I wondered for a moment whether to be afraid. I think I must have remembered some advice someone had given me: if you see a door during your trip, open it. I welcomed the ghouls like children to sit in my skirts and be comforted. Their dementor expressions turned into adorable little sucky faces. I felt an incredible surge of maternal strength.

At some point I became aware that Judith was snoring quietly beside me.

Hey, isn't she supposed to be guiding me? I thought. But then I realized, oh, wait, she has made this space for me to guide myself. Okay, then, let's get on with it. Who do I want to see? First up: my ex-husband. There he was again, the vine around my trunk.

You know what? I said to him. *Actually, I don't want you to be part of my trunk anymore. You can have some space in the roots. Get off of me.* I slowly swept my hands along my torso. *I want my pancreas back now. I'm going to lift it up into the light. I want my heart back, too.* And then I pictured large claws constricting my heart. I saw myself working mightily to pry the stiff talons off one by one.

I channeled Judith, since she was still snoring. *How does this make you feel, Florence?*

It makes me feel like light, but more solid. It makes me feel like marble. It makes me feel like a statue of Athena, but not really, because I can't possibly be that powerful. I am a piece of Athena. I am a piece of her sandal strap! This pleased me greatly. Occasionally I peeked below the eye mask and watched, fascinated, as my fingers grew and shrank and shot out jiggy laser beams. A philodendron plant under the window seemed less real than the negative space between its stems.

Next up: Mom. My dead mother, along with my beloved dead mother-in-law in a surprise cameo, fluttered above me as blue-and-white cloths shimmering in the sun. The mantles sheltered me and sent me their refracted light. It was so beautiful. I cried.

"How are you doing?" Judith had snorted herself awake.

"Great," I said. I didn't know it, but hours had passed. "But can we change up the music?" I asked. Eno sounded like he'd been on repeat for a geological age.

"How about some Hildegard von Bingen?" Judith sounded groggy, but she was coming to. Hildegard was perfect. I'd long been a fan. Not content being a servant to the monks, the twelfth-century nun talked and then protested her way into running her own abbey high in the mountains of the Rhine, where she composed ethereal chants, healed the sick, and wrote lengthy books about medicinal herbs.

With the chants playing, I saw angels dressed in yellow, floating tight like happy sardines in a golden sky. Divine, I was one of them. I cried some more.

At some point, I became a glowing moon. My children drifted about. I was aware that the moon was both whole and still becoming. It was mottled, but it was luminous. There was no shame, no fear. The man I was dating orbited around a bit. I was happy to see him, but I was aware of not really wanting or needing him to get too close. I was doing my moon thing.

Afternoon became evening. Judith asked me if I wanted to go outside and sit in the garden. I was so comfortable, but I also wanted to see some nature. We sat outside and a large banana plant slowly waved to me. The green was *so green*. She brought me a fruit plate. The blueberries and strawberries and crisp almonds were a revelation. I felt peaceful, grateful, content. We ate some broccoli and took a slow walk around the block. I wondered if Judith's neighbors were like, There goes another one.

I was mostly myself, but my limbs didn't feel fully attached to me. In Judith's guest room for the night, I spent another hour or so listening to music. I didn't sleep well and rose the next morning with a headache. As the day went on, I experienced a dramatic come-down, during which I wondered if anything I had perceived with great conviction the day before meant anything at all. The world seemed more full of oppression than ever. I was certain, under the influence of the drugs, that I had seen the truth. Now, I was less sure. In an afternoon session that Judith called "reintegration," we talked about this. Even within my trip, I had reached some conflicting insights. Under the influence of the MDMA only, I recognized that my ex was part of my trunk, that he was in most ways a benign (although still constricting) presence there. During the psilocybin journey, I wanted him off, and saw I must work hard to untwine him. Which was the more real image?

"Okay, sit with that a minute," said Judith. "How are you feeling right now, to talk of letting him go?"

I started crying. "It makes me sad," I said, plucking through her box of Kleenex. "I'm so confused. In my trip, I was so ready to let go, but now I see I'm still really sad."

"It's a process," she said. She told me I still needed to work through my sadness, to feel it, even to welcome it, before I could really let go.

"I'm telling you this," said Judith. "The sadness will go away."

21

OPEN SESAME

Out of the chaos would come bliss.
—DYLAN THOMAS, "BEING BUT MEN"

My Ecstasy hangover lasted for days, during which I regretted having taken such a large dose of it. I looked up my symptoms, and was not pleased to learn that at least one scientist believes that even one hit of MDMA can permanently damage serotonin pathways in the brain. Temporarily, at least, it depletes one's supply of feel-good neurotransmitters. For me, it was several weeks before I felt a normal sense of well-being.

My emotional volatility included some profound moments of optimism. On the flight home, I looked out the window as the plane banked over the Sierras. The view stunned me. Something new was still thrumming in my brain, the old jaded networks not fully back in control. Michael Pollan quotes a neuroscientist who describes psychedelics as shaking the snow globe of your brain; the glitter doesn't necessarily land right back where it used to be. The world outside the plane looked as though it were written in a fancier, cursive script. I was reminded of a line by British journalist and naturalist Michael McCarthy about the "sudden passionate happiness" elicited by the natural world. I didn't want to miss those moments.

I also felt energized and focused, what psychologists sometimes call the "clean-windshield effect"—as if some lint had been vacuumed out of my brain. I hoped I could hang on to that sense.

While I may not have yet achieved the forgiveness-and-friends-forever decoupling expectation during my psychedelic experience, there was another goal, a major one, that I felt was careening closer. It was the one dangled as a possibility by Dacher Keltner and Paula Williams: openness, the personality trait correlated with empathy, creativity, emotional richness, imagination, intelligence, and intuition. It's the jackpot trait: one Australian study showed that a one-deviation jump in openness is equal to a $62,000 gain in personal income in terms of life satisfaction and happiness.

This trait is of great interest to the psychedelics world, especially since researchers at Johns Hopkins University School of Medicine published a notable paper in 2011. As part of their ongoing quest to understand the effects of these drugs, investigators analyzed personality questionnaires—the 240-item NEO—from two double-blind controlled studies, in 2006 and 2011. Fifty-two volunteers filled out the questionnaires before and after receiving either high or low doses of psilocybin or a placebo. In the NEO, they have to rate such openness statements as "I have a very active imagination," "I am intrigued by patterns I find in art and nature," "I experience a wide range of emotions and feelings," "I often enjoy playing with theories or abstract ideas," "I consider myself broad-minded and tolerant of other people's lifestyles," and "I think it's interesting to learn and develop new hobbies."

It should be noted that anyone who would willingly sign on to a magic-mushroom study is likely to be fairly open-minded to begin with.

Even so, those given the high dose of psilocybin—but not those in the other two groups—became markedly more so, and remained that way a year later. Their "openness" scores jumped from 66 to 72, enough to lift them from the category of being pretty open to being mega-open. (They showed no changes in the other four personality domains of neuroticism, extraversion, conscientiousness, and agreeableness.)

This was a big deal, because personality traits are believed to remain relatively constant over time, like your IQ or your love of pistachio ice cream. If anything, openness is supposed to decrease a bit as we age. Each decade makes us (on average) a little less curious and broad-minded.

The authors believed this to be one of the first studies demonstrating changes in personality in healthy adults due to an experimental event. The Hopkins research spawned some interesting follow-up work. In one study from 2017, more than 800 people provided information about their lifetime experiences with psychedelics, cocaine, and alcohol. People who had experienced "ego dissolution"—such as when I felt my "self" to be indistinguishable from other filaments of light or the beads in the curtain of molecules—were more likely to report holding liberal political views and feeling connected to nature, and they were found to have higher openness scores.

If changing your openness trait can alter your personality, why not try it in people who really need a personality change? Or those who could use a return to their original, sunnier personalities, for example, people suffering from post-traumatic stress? For some people, experiencing a catastrophic event can alter their personalities in destructive ways. The full spectrum of PTSD symptoms includes hyperarousal, hostility, mistrust, social withdrawal, hopelessness, impulsivity, and low self-esteem, according to the *International Statistical Classification of Diseases and Related Health Problems*, making normal functioning difficult in many areas of life.

In a 2017 study at the University of South Carolina, neurologist Mark Wagner and colleagues assigned 20 people who had experienced emotional trauma through war or violent crime to undergo either two sessions of MDMA-assisted psychotherapy or two sessions of psychotherapy plus a placebo. Both groups improved, but the MDMA group experienced significantly bigger drops in PTSD symptoms, and the drop corresponded to increases in openness on the personality questionnaire. Those improvements remained even three to four years after treatment. "These findings are consistent with the notion that increased openness may be a mechanism of therapeutic change," wrote the study authors.

And what was facilitating the openness? As suggested by Williams, Keltner, and others: awe. In the Hopkins study, the researchers noted that the higher the participants rated their experience of mysticism, the more likely they were to experience increases in openness. About two-thirds of the high-dose takers met the criteria for having had a complete mystical experience. Hopkins researchers now routinely measure the degree of mysticism that their participants encounter. In over 600 subjects, 60–80 percent of them report a very profound, cosmic adventure.

Naturally, I was curious to know if I, too, had lucked into a "complete mystical experience." When I told a friend I would take a questionnaire to find out, he rolled his eyes.

"The mystical is ineffable," he said.

But is it really? Mystical experiences recorded through time, at least as described by spiritual thinkers across many religions, share several key elements. Among them are an overpowering sense of unity, or the interconnectedness of life; the preciousness or sacredness of the experiences, something impressively deserving of respect; what William James identified as a noetic sense, that the insights gleaned feel undeniably "true"; the openheartedness and joy of the journey; and the transcendence of time and space.

The Hopkins team created a validated questionnaire, repurposing an older survey, the "Mystical Experience Questionnaire" from the 1960s. They kept the name, which they later admitted was a bit of a branding error because it sounds more woo-woo than it is.

I dove into the 30-item mysticism questionnaire. I had to rate on a scale of 1 to 5 such statements as whether I experienced "the fusion of your personal self into a larger whole" (yes, a lot!), a "sense of being at a spiritual height" (yes!), a "sense of awe or awesomeness" (yep), and a "feeling that you experienced eternity" (okay, that one does seem a little ineffable). There was that moment, after all, when I was listening to Hildegard and saw those angel-sardines floating together in the sky. *Wait,* I thought at the time, *we're not seeing God, are we? Because I don't really go in for that.* And the answer came back loud and clear: *We are all God.*

Hildegard, it's worth pointing out, had mystical visions of her own, beginning in early childhood, perhaps occasioned by migraines. They were so powerful that she meditated on them the rest of her life. She wrote about a recurring theme: "In this vision my soul, as God would have it, rises up high into the vault of heaven and into the changing sky and spreads itself out among different peoples, although they are far away from me in distant lands and places. And because I see them this way in my soul, I observe them in accord with the shifting of clouds and other created things." Classic! The unity, the cosmic quality, the disintegration of self. Although largely feminine in nature, her beatific visions were accepted and eventually celebrated by the Church establishment. Kings and queens—Henry II and Eleanor of Aquitaine—considered her a spiritual adviser.

I loved how feminine my own visualizations had been. The angels were clearly gendered. There were mothers everywhere: I, in my skirts, harboring the ghouls; my own mother and mother-in-law, the two of them embodied as gently billowing sheets of cloth in the sky, creating a bower of shelter, light, energy.

It will not sound surprising, then, that according to my questionnaire results, I did indeed have a complete mystical experience. One more step toward awe-and-openness.

"I think awe really helps us experience transcendence," said Peter Hendricks, a clinical psychologist at the University of Alabama, who in

2017 wrote a paper titled "Awe: A Putative Mechanism Underlying the Effects of Classic Psychedelic-Assisted Psychotherapy." When I called him to see what inspired him to turn his spotlight on awe, he said the emotion is still overlooked as a central element in the power of psychedelics to create transformation and transcendence.

Transcendence was a fuzzy concept to me, not one that I'd ever spent five minutes thinking about. What did it mean? Hendricks tried to explain. "To be the best versions of ourselves you need to transcend the self and dedicate your life to something bigger than you," he said. "Often you can achieve a recognition of that through peak experiences, and those are typified by awe. Awe ideally makes us realize that all humans are in it together, all creation is in it together."

In fact, "awe may be the quintessential binding emotion that drives social integration and cooperation," Hendricks has written. "If evolution ultimately selects for cooperation, it follows that awe represents the pinnacle of human experience."

That's a pretty grandiose claim for an overlooked emotion. Humans used to run smack into awe naturally through our once daily but now lost encounters with the Milky Way, the sunset, the phases of the moon, and wild animals, both friendly and fierce. The human brain is wired to feel awe. If this emotion represents the pinnacle of human experience, we are now greatly impoverished.

That feeling of collective destiny, I was learning, is important not only for the future of civilization but may be the best defense we as individuals have against loneliness. It is increasingly recognized by psychologists as being critical for well-being. It's what got many through the darkest days of the pandemic.

In 1945, Abraham Maslow, the president of the American Psychological Association, conceptualized the famous hierarchy of needs: Physiological (food, water, shelter), Safety (security through rules and laws), Belongingness and Love (social affiliation), Esteem (recognition or achievement), and Self-Actualization (fulfillment of personal potential).

What is less known is that in 1969 he amended the schema to add a sixth, higher, level: Self-Transcendence, which he described as "a cause beyond the self" and "a communion beyond the boundaries of the self through peak experience." Transcendence, he believed, wasn't likely to just happen on its own; it arises through extraordinary events.

Maslow, it turns out, was a bit of a mystic. I wondered if he'd taken psychedelics. Plenty of psychologists in the 1960s did. In the famous Marsh Chapel Experiment from 1962 conducted by Harvard's Walter Pahnke and Timothy Leary, 20 divinity students from around Boston took either psilocybin or a placebo. Eight of the ten on psilocybin described a profound mystical experience in line with the classic reports in religious literature—and with my own, on Judith's futon. One subject, Huston Smith, who would become an esteemed religious-studies scholar, later said that day was "the most powerful cosmic homecoming I have ever experienced." The study was criticized for lax protocols, including the fact that the lead investigators were likely dosing themselves, according to Johns Hopkins psychopharmacologist Roland Griffiths. Perhaps that is why Griffiths was motivated to replicate the study, which he did in 2002, finding similar results.

Bill Richards, a psychologist at Hopkins and a frequent collaborator of Griffiths, studied under Maslow and knew him well. Richards said that while a heart condition kept Maslow away from psychedelics, he nevertheless described such peak experiences as lying naked in the sun in his backyard. (That doesn't seem like such an extraordinary event, which is cause for celebration. Some people have, or can cultivate, an ability to be exceptionally tuned in during an ordinary day.) Whatever he was on, I'd like some.

HENDRICKS ARGUES THAT the defining characteristic of awe may be the diminishing of the self at the altar of something much larger. Psychedelics facilitate this through scrambling up our normal neural connections. Brain-scan studies of people tripping on mushrooms show a drop in activity in regions like the posterior cingulate cortex (PCC) and the medial prefrontal cortex, which play a big part in self-awareness and self-control. The drugs also disrupt the pathways between our thinking brains and deeper structures tied to emotions and memories. Known as the default mode network, these linkages enable us to place our egos at the center of how we think and relate and move through the day. They help us get along with others, read social cues, be introspective, and understand who we are. But too much self-referential thought can distort our emotions and hamper creativity. Psychedelics offer a chance to observe what a shaky drama queen our perceiving self can be.

Mindfulness meditation also diminishes activity in this network, as can experiencing an arresting piece of art or moments in nature. Iris Murdoch writes of a beautifully mundane, nonpsychedelic encounter with a flying kestrel. "In a moment everything is altered. The brooding self with its hurt sanity has disappeared. There is nothing now but kestrel. And when I return to thinking of the other matter it seems less important." She calls this process "unselfing," and we clearly don't do it enough.

When we experience less *me*, we experience more *us*. "Awe," said Hendricks, "is where attention gets drawn away from yourself in that you don't notice you're there. There's something else going on that's so big and so different, and there's healing in that." His clinical work largely focuses on chronic pain, rumination, and catastrophizing.

He knows something about catastrophizing. Hendricks shared with me that he, too, was recently divorced. It had been two years. He had three young children. The split, he said, was far worse than anything he could have anticipated. "I lost my father in 2001 and I was able to bounce back from that, but for me, this lingers. I live with it on a daily basis."

"Has awe helped you?" I asked.

"Yes," he said, explaining that he thinks it can deflect or defuse the nagging thoughts that come with a negative experience. "With divorce, you ruminate. Things will never be the same; I'll always long for that relationship; I'll be damaged and carry stigma and failure and the way it's impacted my children. So when you can get out of circular rumination, it can be really helpful. Awe helps you see things from a different place. Maybe you can return and say, yeah, I'm heartbroken, but maybe it wasn't good and I can learn something. Maybe I can approach things more creatively and optimistically. This is what we call post-traumatic growth."

WHEN I CAME home, I described to my therapist, Julia, the alternate visions I'd seen of my husband being part of my trunk (under the early influence of the MDMA) and wanting him to get the hell off of me (under the psilocybin). She looked pleased by what she considered progress.

"It sounds like you are decathecting," she said. *Cathect*, I learned, is an obscure word used almost exclusively in psychoanalysis. It derives from the Greek word *kathexis*, for "holding, possession, retention." It probably didn't enter English until the twentieth century, when it was popularized by Freud—or rather, by his devoted English student and translator, the charismatic psychoanalyst James Strachey. To Freud, writing *Mourning and Melancholia* in 1917, cathexis meant investing a beloved person or idea with powerful psychic and sexual energy and attachment, so much so that the object becomes bound with the lover's sense of ego. (Months after my conversation with Julia, I was excited to discover the word in a paper my mother wrote and delivered at the American Psychological Association convention in 1973, when I was six years old. I wish she were still around to explain it to me.)

Julia said the untwining—wrenching though it may be—is a necessary stage of grief and mourning after the loss of love. The Freudians would say the ego can then reattach the freed libido to a new object, although often, as Freud later acknowledged, this process is imperfect.

As the word's English ambassador, Strachey knew a thing or two about heartbreak. A peripheral member of the Bloomsbury Set, he fell in love at Cambridge with the poet Rupert Brooke, who did not reciprocate. The poet's death at age 27 in 1915 from an infection acquired in Greece during World War I left Strachey shattered. Strachey was himself pursued by John Maynard Keynes and by famed mountaineer George Mallory, who in turn was lusted after by James's brother Lytton. It's hard to keep track of it all. Lytton wrote to Vanessa Bell of his muscular, athletic crush: "Mon dieu! George Mallory! My hand trembles, my heart palpitates, my whole being swoons away at the words—oh, heavens!"

Alas, poor Lytton's infatuation was not returned. Devastated after Mallory departed from a visit, he wrote to his brother: "The sunshine has gone out of my life. . . . I'm oppressed by the agony of human relationships. It's not only the love affairs that are bound to fail! And now I shall never see him again, or if I do it'll be an unrecognizable middle-aged mediocrity, fluttering between wind and water, probably wearing a timber toe." A wooden digit would have been preferable to Mallory's actual fate: he died making the first attempt to summit Everest in 1924.

But I digress.

Facing my own middle-aged mediocrity, I was finally ready to de-cathect from my former love. I was also ready to de-cathect from my sense of unworthiness. On the river trip, I'd come face-to-face with my flaws in the shadowless desert. Now, as with the ghouls, I was ready to welcome them, even laugh at them. It felt incredibly liberating after a lifetime of trying to be good at everything.

But I still wasn't ready to cathect with another. Julia and I discussed my hallucinatory vision of not wanting to share my planet with anyone else, not yet. (And yes, it felt goofy and somewhat inane to be talking about relationships this way, *my planet*, but there it was. It was working for me.) I also recognized something new was starting to happen, and it wasn't just my embrace of Age of Aquarius vocabulary: I wasn't so afraid of the unknown anymore.

22

MAN IN THE KASTLE

Opioids, Love, and the Science of Recovery

Straightway she cast into the wine of which
they were drinking a drug to quiet all pain and
strife, and bring forgetfulness of every ill.
—HOMER, *THE ODYSSEY*

As I settled into my routines after Portland, my diabetes markers stabilized and even slightly improved. Was it that I'd envisioned releasing my pancreas into the light? Or the fact that I'd become very good at eating like a hamster and exercising after meals? Was I, finally, calmer and releasing less noradrenaline? For now, at least, I could put off the need for insulin, and my endocrinologist wanted to give me a medal. I'd managed to regain a few pounds. I was sleeping better, crying less, not smoking weed, and, of course, hiking all over the parks of my city, often with friends, kids, or my father. Dad and I frequented a trail in Rock Creek Park that was about halfway between our homes. We walked there regularly as the beech and oak leaves turned golden and pink, then fell off altogether. Every week the light changed, as did the sound of the crunch underfoot.

Small, frequent doses of nature—not epic ones—were proving both calming and energizing. I caught up with Steve Cole and told him all my news.

"Good," he said. "Send me another blood sample."

I promised I would.

But first, for the sake of my transcription factors, I needed to leave one of the most cynical places on earth—Washington, DC—and go find some social optimism.

I LIKE BOURBON and the greenest of green horse pastures, so in early spring I headed to Lexington, Kentucky, home to Nathan DeWall, a neuroscientist specializing in social exclusion. DeWall, who works in the brick Federalist-style Kastle Hall on the University of Kentucky campus, is known among his peers for having kooky ideas that pay out. He came up with the experiment giving Tylenol to people who feel rejected and then scanning their brains (yes, it helped them feel better). Because Tylenol binds to cannabinoid receptors in the brain, DeWall next looked at marijuana (it also helped, but not as consistently, and not without side effects like increased irritability, envy, aggression, and misanthropy, not to mention, probably, the munchies). "Our findings do not advocate for widespread use of marijuana to treat all forms of emotional pain," he has written. "Although socially disconnected individuals may smoke marijuana as a self-medication for the adverse effects of their social situation, there are likely more efficacious therapies that are not associated with the health risks of regularly smoking marijuana."

Thin haired and pale, DeWall exudes a warm, nerdy curiosity. Wearing a blue golf shirt and surrounded by four computer monitors and multiple sets of spectacles, he seems an unlikely lead singer for his classic-rock cover band, Roar Shock (get it?). Like so many of the researchers I spoke with, he is no stranger to emotional pain. A few years ago, his work was featured on a radio show, and his mother called him up. "I'm so proud of you," she told him from her home in rural Nebraska. "That was the last time I ever talked to her," he told me. Soon after, she was walking in her driveway, fell, and hit her head. "I got the call in the middle of the night, and then she went home to God," he said. "That was social pain like I had never experienced in my life, like getting hit in the gut. I gained weight. It changed my life." Two years later, another blow: the baby he and his wife had been in the process of adopting was reclaimed by her birth mother.

DeWall dealt with his sorrow by running. He decided to run the longest, most difficult ultramarathons in the world: the 135-mile Badwater in Death Valley, and the 147-mile Marathon des Sables in Morocco. Not only were these races bloody long, they were miserably hot. You needed to prepare mind, body, and soul for these sufferfests. The runs helped in two major ways, he told me: First, they are social. The races are so hard that you can't do them alone, forcing you to bond with the other runners, with your training partners, and with your support teams. Second, they require preparation and logistical planning on par with Caesar's advance army. DeWall knew from brain scans that social pain and cognitive effort have an inverse relationship: if your frontal cortex (thinking brain) engages, you feel less pain in the emotional centers.

"Whoa," I said, describing my 30-day trip to him, "do you think all those logistics were helpful?"

"Maybe!" he said. There's not a lot of data on grief and expeditionary planning. What really gets him excited these days is not measuring the mental-health effects of endurance sports, or marijuana, but another burgeoning form of self-medication that appears tied to the epidemic of loneliness: opioids. Lexington lies in the heart of Appalachia, where unemployment has been relatively high for years and where there is limited access to healthcare. The mortality rate there from overdosing on drugs is 70 percent higher than the rest of the US. In Kentucky alone, about a thousand people per year die from opioids, including heroin, prescription drugs, and synthetic fentanyl.

Curious about the epidemic, DeWall asked a local acquaintance if she knew anyone addicted to opioids. She told him she knew six people who had died. "I was haunted by that conversation," he said. It didn't make sense to him that so much policy discussion revolves around cutting off the supply of habit-forming drugs like Oxycontin. "What about the demand? Who are the people who tend to use and abuse? Do they feel they are at the edge of society, with no job, no voice? They are a paragon of social disconnection." He believed addiction is, ultimately, a disorder of attachment.

So DeWall decided to run a pilot study. He transported opioid addicts to the lab at the university and ran them through fMRIs while they played the classic rejection game, Cyberball, in which they get left out of the ball toss, and then viewed pictures of marijuana, random pills, opioids, and neutral images. He wanted to know if the pain of rejection increases the craving for drugs in the brain.

He found that, per usual, the subjects' DACC—the dorsal anterior cingulate cortex—lit up when the subjects felt rejected. That was their pain signature. Then, when they saw the pictures of the opioids, but not the other substances, DeWall saw fireworks going off in the caudate nucleus, which is a part of the brain associated with craving and reward. Interestingly, it's also what lights up in other studies when pain sufferers look at pictures of loved ones. The opioid users were, said DeWall, "replacing people with drugs."

This makes sense when you learn how opioid receptors are key players in our brain's social attachment networks. We need to be motivated to pursue human relationships in order to mate, raise young, and build cooperative societies. When you remove a baby monkey from its mother, the infant's natural opioids plummet, driving it to seek Mom to feel good again.

On the other hand, when you give morphine to monkeys, they instantly become less interested in their mothers, in grooming, and in mating. Their opioid receptors are already satisfied. Such were the observations of one of DeWall's academic heroes, the late pioneering social psychologist Jaak Panksepp. Trained in the Skinnerian tradition that dominated the field in the mid- to late twentieth century, Panksepp and others once believed that animal behavior, including ours, was largely determined by innate responses to environmental cues. Morsel of fruit: hunger: cue eating. Mom provides food: cue "love." Emotionality was considered relatively unimportant, plus it was hard to measure and too subjective.

But in 1991, Panksepp's teenage daughter was killed by a drunk driver on a dismal Good Friday evening near his home in Bowling Green, Ohio. As he slowly began to recover from the crippling grief, it was with a renewed determination to give emotions their due. Panksepp was adept at finding animal models to study such behaviors as mating, parenting, play, and avoidance. He famously tickled lab rats and recorded their laughter. He gave ADHD drugs to rats and watched them stop playing and exploring their environments. As neuroscience grew more sophisticated, he found hormonal and neurological pathways for core emotions and behaviors. As for opioids, he wrote, they "have powerful influences over our feelings, especially our negative responses to social isolation."

Panksepp wrote about the French artist Jean Cocteau, an opium addict who recollected in his diary that opium liberated him "from visits and people sitting round in circles." Panksepp recounted Homer's description of the reunion of men returned from the Trojan War in *The Odyssey*, remembering their fallen compatriots with a "twinging ache of grief," for whom Helen's special wine served as "an anodyne, mild magic of forgetfulness." He believed Homer was referring to a tincture of opium or cannabis that activated the brain's large synaptic opiate receptors, among the brain's most powerful modulators of pain.

It's because we are built for love that we become addicted to other, less sanguine things. And opiate addiction in humans, Panksepp argued, is most common in "environments where social isolation and alienation are endemic." Addiction is the ultimate pathology of heartbreak.

So what do you do about it? Nathan DeWall agrees with Steve Cole that mission and purpose are key elements for treating loneliness, but, at least with addicts, you also have to help them rewire the opiate system in the brain. You must put social connection back into the picture, in a meaningful way. If the craving impulse was designed for attachment, then love is necessary to heal addiction.

DeWall is excited by models like the Recovery Café, a network of nearly two dozen community spaces around the country. The cafés operate on a membership model built around the concept of creating a space of belonging, and peer accountability. Among other things, he explained, the idea is you have people to talk to, to share your story with, and to help destigmatize the shame that comes with social pain and trauma.

"What I wish," said DeWall, as I was preparing to leave his office, "is that people would just talk about this stuff. If you walked in here on crutches, I'd be like, What happened and how can I help you? But with heartbreak, people often just suffer alone. Our mind and body can be resilient, but extreme social pain is the most significant experience you can have, and it lasts a long time."

A FEW MONTHS later, I found myself in Seattle, home of two Recovery Cafés. I headed for one in a large, bright space just north of downtown. Founded by pastor Killian Noe and University of Washington psychologist Ruby Takushi, this was the first one in the country, and it's been going strong since 2004. Since it opened, this café has served more than 300,000 meals to about 9,000 members.

No one in Seattle is far from coffee, and here the serve-yourself counter greets visitors immediately. Beyond that sits a food station for sandwiches, main courses, and salads. You can come as a guest, or if you're a member, you can eat two free meals a day, taking them to an array of small and large tables set up beneath red and orange walls bearing slogans like "Grow," "Dream," "Faith," "Grace." Adult lunch-goers of mixed age, race, and tattoo status sat at most of the tables. A few older men with beards lounged in overstuffed chairs near the window, and a younger man sat at a workstation with a computer. Bookshelves lined the back wall. A few classrooms and offices surrounded the dining area, and a chess set lay open nearby. I made a mint tea and sat at a table to wait for Noe. Soon I was joined by a middle-aged woman carrying a plate of chicken chile. Redheaded, wearing a beige turtleneck and orange lipstick, she introduced herself as Jennifer. She said she had been a member for seven years. A gong sounded toward the back of the room, followed by an announcement about "fresh start" classes every Wednesday at 2 p.m.

"I come whenever there's a class I like," said Jennifer. "I can't come if I'm not sober for 24 hours. That's a rule. I like the art classes and the AA classes. I don't want to be home in the morning with no structure. It's better that I accumulate days of sobriety." Another gong went off and someone announced a birthday, which was followed by singing.

Killian Noe jostled over to greet us both with a big smile, apologizing for running late. She looked surprisingly lighthearted for someone who has been juggling funders, volunteers, and people facing much suffering. Simply dressed, with an athletic build and a graying strawberry-blond ponytail, she exuded both gentleness and rebar-like determination. We settled in for a long chat, during which she explained the basics. This café has about 350 members at a time, hoping to serve them in a holistic, long-term way. There are four pillars to the program: emotional support (including 12-step programs for opioid and alcohol addiction), informational support such as vocational training, support navigating housing options, and "affiliational" support to build social skills and connection.

"The whole concept is all of us need authentic community," said Noe. She told me she grew up in a small town in North Carolina, where family friends disappeared after her parents divorced. It was intensely painful.

"Here, there's a lot of connecting going on." As a young pastor in Washington, DC, 20 years ago, she was struck by the pervasive problems of both homelessness and addiction, and also by the fact that the standard solution—incarceration—didn't work. There was little effort to address the primary causes of addiction: social marginalization, poor support services, and inadequate healthcare. According to the National Institutes of Health, at least two years of treatment is necessary to maintain long-term recovery, and yet most addiction programs in America—when they're available—last 30 to 90 days.

Working in small peer-led groups in DC and then here, she saw people's self-confidence, health, and prospects slowly improve over time. Most of those who come through these doors have suffered trauma, she explained. "Nine out of ten," she said, "but we believe it's more like ten out of ten. That often involves shame, and that's what lies underneath addictive behavior. So we welcome people. Most people here have never been a member of anything. That has positive meaning for them."

In addition to being sober when they arrive and agreeing to volunteer for chores, each member must commit to participating in a weekly "recovery circle" of about seven people, what Noe calls "a loving accountability group" that forms the core of the program.

"When we're trying to remove destructive behavior, we need to replace it with something positive," she said. "Underneath it all is the notion that we all were created for connection, and that loneliness really is the killer."

The café's multipart strategy isn't a fast cure, but it can be effective. One survey found that 92 percent of members feel increased hope in their lives. Many stay sober, find jobs, and return to keep volunteering. Noe introduced me to Michelle James, who spent 10 years in prison and was now working in the Seahawks stadium and coordinating a free women's nature retreat. At 52, she was a bombshell, decked with glossy curls, pink lipstick, and a fantastic velvet jacket. "Believe me, I didn't look like this the first time I walked in here," she said, as we both admired her fingernails. Back then, she was homeless, a heavy user of meth, alcohol, and heroin.

When she joined the Recovery Café "there was no judging," she said. "We talked about things around grief, dealing with pain and being okay in my own skin. It was an open and trusting place to share my story and there was healing."

James has been clean since 2016, but it's still difficult. "I'm often running myself down and I begin to see depression coming on. The fuckits, I call it," she said. What's helping her now, in a light-handed way, is exactly what Cole, DeWall, and Williams tell us will help: belonging, working on behalf of others, and finding beauty.

From across the table, I found myself gazing at her necklace. It was a little hard to make out. "Is that a butterfly?" I asked, thinking back to the visions of the children in Joplin during the tornado. For them, the butterflies meant protection, but the symbolism of passage and emergence was hard to ignore.

"Yes! I have an assortment I've been collecting," she said, touching it gently. "I hadn't seen any in years and now I always see them."

23

THE FUTURE OF HEARTBREAK

Your great mistake is to act
the drama as if you were alone.
—DAVID WHYTE,
"EVERYTHING IS WAITING FOR YOU"

It's a far distance between the Recovery Café in Seattle and the café of the Museum of Broken Relationships in Zagreb, Croatia. And yet the two are closer than they appear. In both, wounded attachments are salved by strong caffeine and the presence of others similarly afflicted at one time or another.

It had been a year since my river trip, two since the marriage's real sundering. I'd been asked to give a summer talk in Slovenia about the human connection to nature. Not much of a fee, but all expenses paid. I had enough frequent-flier miles to bring my daughter, who thought Ljubljana looked like Paris on Instagram, so she was game. And it did look like Paris, but it was less crowded and cheaper. We ate at riverside cafés, learned about resistance art, and spent two days in the mountains canoeing and identifying invasive species as part of a "bioblitz" program sponsored by the US embassy and the National Geographic Society. Now 15, my daughter was, like me, a city and nature child, and therefore my ideal traveling companion. Plus she was willing to let me sample her outrageous eastern European pastries.

She didn't know much about the book I was writing, and it was time to tell her. While not examining rust fungus, we'd both been reading the Brontë sisters, for whom heartbreak was less about rupture and more about the agony of unattainability. Love was close enough to touch but not to have; if you got it, briefly, you had to either die, go blind, or burn a lot of stuff down. No simple trick, this marriage plot.

Would she be willing to travel two hours by train to Zagreb to see the Museum of Broken Relationships?

"Can we go to some thrift stores?"

THE MUSEUM OPENED in 2006 as a lark by a pair of (split-up) artists, Olinka Vištica and Dražen Grubišić. A collaborative and interactive concept, the museum has become one of the top tourist spots in Zagreb. It now occupies an elegant mansion in a hilly neighborhood just down a narrow street from St. Mark's Church and a plaque commemorating Nikola Tesla.

Grubišić met me in the bright café while my daughter wandered around the displays of objects donated from around the world by the jilted, the aggrieved, the sorrowful, the sardonic, and the indignant. A thin, goateed video producer in his 40s, Grubišić ordered an espresso and explained how he got here.

"It's funny because the idea was not to do this for visitors," he said. "It was meant to give people a kind of ritual to make it easier for them when they break up. That was the main goal. We'd just keep things in dusty hallways." It started when he and Vištica broke up after a four-year love affair. They couldn't figure out what to do with the cheap toy bunny that he had once given her and that seemed to carry so much emotional weight. He thought it might be redemptive to tell its story. "When you can distance yourself enough to write about it, it helps you. For a long time, you might not be ready, and then maybe you are ready."

Heartbreak, they decided, was in need of ritual. "It's such a serious thing and there's nothing to help with resolution," he said, stirring his small drink. "For guys it's even worse. You're not even supposed to mention it, or you just say, we broke up, and then you go out and get drunk. And it's not something you're used to. It doesn't happen like once a month. Sometimes people experience it maybe twice in a lifetime. How do you cope with this? I mean even death is more common than breakup, in a way."

Grubišić believes that heartbreak deserves not only institutional respect but the space for reflection. The museum provides this, although more than anything, it helps codify these wounds as a shared, universal experience. But it is the particulars of each grief—often odd and highly eccentric and worthy of a quick story—that help create the psychological distancing required for healing. The museum's donors can now shake their heads and say, Can you believe this story of why I knit a misshapen sweater? Or of this water bottle shaped like the Virgin Mary or of this espresso machine? The artifacts, like the memories they convey, become both enshrined and released at the same time.

Grubišić rose to show me around. There's a system to the curation, he explained, with more humorous items displayed in the front exhibit rooms. The objects lay carefully arrayed on pedestals and in well-lit cases as if they are precious art. An interpretive text, written by the donor, accompanies each. Pizza-dough mix, bicycle tools, a crusty scab from a boyfriend's leg in Austria. "I kept one of his scabs after it had fallen off, with the (not so serious) idea in mind of having him cloned in the future if need be. . . . In the end, my constant fear for him led to our breakup. . . . I have kept the scab to this day, for twenty-seven years." Someone from

Finland shipped in an exercise bike. "When I found out that my dear wife liked to ride much more than just the exercise bike, I divorced her."

I asked Grubišić about the cultural differences in the way people describe the objects they send into Croatia and to traveling exhibits in many different countries. He smiled and nodded. "In Asia," he said, "you get most of the stories ending up with some sort of gratitude, like we enjoyed each other and so on and so on. The US is incredibly self-centered. So many of the stories are how I'm absolutely perfect and he or she is horrible and did this and that." Sometimes the objects are very much of a place. In Los Angeles, he recalled, someone sent in a silicone breast implant her boyfriend insisted she get (the implant, like the man, later got expelled). In Croatia, a veteran sent a prosthetic limb he received from a once-beloved social worker. The French are much more cerebral in their accompanying notes, making philosophical, well-argued cases. The curators have received more objects than they can handle from southern Europe and Mexico, where there's more "emotional sharing," he said.

Many of the objects come from short-term relationships. The Virgin Mary water bottle: "He was discovering Europe by train. We met at the Buddha disco." A boxy radio: "I was given this radio at a beach in Rijeka in 1984 by a guy named Darko . . . It remains quiet because of the broken relationship it has come to represent."

The displays are, like the pain of heartbreak itself, both reverential and, at times, absurd and banal. Here heartbreak finally finds a theater, a full expression of its performative yearnings. And yet, wandering around as viewer, I was struck by how the stories are, like dreams, often untrustworthy. Every relationship starts as a collaboration, but each one ends with its own unreliable narrator. These details of grief don't offer explication so much as supplication—see how it was before it broke me, see how I suffer still. Here was more evidence that many people struggle to move on. One long-married woman sent in a tattered stuffed animal. "He gave me Snoopy on my 17th birthday. . . . He fell in love with another woman and he chose her. . . . Telling me he hadn't loved me at all in those 30 years. I just don't understand."

By now I respected the wounds of heartbreak and knew its labilities firsthand. I also longed to put my story to rest in a case with a label and a four-sentence narrative. But the museum's contributions also show that heartbreak is not a condition to be bandaged and cured. It is more like a bruise in the brain. It never really disappears. It leaves us altered. If we're lucky, new, stronger neurons and pathways grow around the sore spots. "Heartbreak is how we mature," says the poet David Whyte.

If human connection lies at the root of heartbreak, it is also required for its transformation. It's no accident the museum started in Croatia, which has seen in recent memory the violent, political rending of a nation, where friend turned on friend and disunion tore culture and history apart. Grubišić was in his early 20s during the sectarian wars following the breakup of Yugoslavia. "With so much separatism and violence," he said, "that's part of what attracts me to such a unifying emotion."

It was interesting to hear him call heartbreak, which I'd once thought of as one of the most deeply isolating experiences you could have, a unifying emotion. If the past was personal, he was saying, the future is both personal and collective.

"We are all the same," said Grubišić. With that, he said goodbye and left me at the gift shop.

The place was packed. Young visitors, old visitors, some alone, some in small groups, speaking different languages. It was a small river of people who had been touched, seared, maybe shattered by love, yet somehow patched back together. A T-shirt splayed on a hanger in front of me. "You're My Everything. Not." I barked out a laugh and then walked on to find my daughter.

24

THE PERSONALITY OF THE BODY

The body is not a thing, it is a situation: it is our
grasp on the world and our sketch of our project.
—SIMONE DE BEAUVOIR, *THE SECOND SEX*

Before I left for Croatia, I'd sent that later blood sample to Steve Cole. Now he could compare biomarkers across three time points: some months before the river trip, right after it, and then nine months later, two years out from the breakup. Rather than just spotlight his favorite 53 immune-regulating genes, he had decided to widen the lens. We each have 20,000 genes, and the ways they get expressed tend to stay fairly stable from month to month. Cole was curious to see which of mine had changed significantly, and in which direction.

It had taken a while to get the analysis performed because Cole waited to process my vials of blood with a larger batch for a study he was running. We met over a couple of calls, the last one during the spring of 2020.

Like so many conversations during those pandemic months, it was held on a video platform. He was happy to duck into a quiet spot away from his kids, and he was growing a quarantine beard. "I cut it weekly, as if it were a lawn on my head," he said. He looked less golfer and more Jung. It suited him. As someone who studies loneliness and despair, he was now understandably concerned about the short- and long-term health effects of the Covid-19 pandemic and the lockdowns on so many people throughout the world. He and his colleagues would have years of work ahead of them sorting out just how bad it was.

For now, he had some time to ponder the cellular fingerprints of heartbreak. He was more excited than he'd been with the ho-hum results after the river trip. He was pleased with this new, more global approach to the analysis, looking at changes in transcription factors across the whole genome. He was looking for signals in a storm. "There are small subsets of genes that respond to your life," he said. "But which and what? If we know something about them, we can translate that into various biological lightbulbs."

And now, some good lights were shining. "So, at Time Three," he said, "you're looking pretty good."

It was hard to know how my gene expression compared to other people's, for example, those of the happily paired-off, because not many people have been analyzed for these markers, and among those who have, Cole is reluctant to compare one batch of samples to another. So these results are within-subject: how I looked compared to myself over time.

Here's what changed: The so-called CREB family of stress-related genes was less active, indicating I was producing less norepinephrine at Time Three than at Times One and Two, and as a result my genes related to fighting viruses were likely looking much better. "By Time Three, you were really doing pretty well in terms of antiviral biology," he said, noting that I was, for example, making more interferon-related factors that aid immune cells. "If it makes you feel any better," he continued, "the leukocytes are on your side." (It made me feel better. Remember: pandemic.)

Also, he said, my blood showed less expression in inflammation-related genes, although I still had some notable levels probably due to having autoimmune-related early-stage diabetes. Still, the momentum was now heading in the right direction.

"Your transcription factor signatures look like less inflammation and more antiviral response," he said, explaining that he looked for markers known to influence specific immune cells. In particular, I was generating more transcription factors known to buoy type 3 dendritic cells. I asked him to explain.

"So dendritic cells are these little cells that patrol your body looking for trouble," Cole said. "You can think of them as highly mobile reconnaissance cells." Or like spies. They take samples of the invader virus back to the lymph nodes and tack them up like a Wanted poster. The lymph nodes are a convention of other immune cells like T cells and B cells that read the poster. Ideally, a few of them say, hey, I recognize that one and I know how to disarm it, and off they go. As people age—and when they are lonely—they make less of the really effective, virus-killing type I interferon that each cell needs to repel invaders like a door bouncer, which is why we want good dendritic cells, he said.

As Cole explained it, kids and young adults likely weren't getting as sick from Covid-19 because they have great first-line-of-defense interferon responses. Some older people got so overwhelmed by the invading coronavirus that their bodies had no choice but to launch full-on cytokine storms for an inflammation defense, but that response caused widespread organ damage. Someday, it might be possible to look back at the blood work and the transcription factors of those who succumbed. Cole wasn't so concerned about several months of social isolation for those in lockdown. What was distressing him now were the newly unemployed, people without hope and without safety nets. The gene expression of the involuntarily unemployed looks even worse than that of lonely people, increasing their all-cause risk of death 70 percent during the first year, more than double what it is for the newly divorced.

Cole wasn't sure how I'd do if I developed Covid-19 in the pre-vaccine time in which we spoke, but for now, he was reassuring. "Your body doesn't look like a person who is fundamentally deeply threatened or rattled," he said. (My sample was pre-pandemic blood, though, so it might indeed have regained some rattle by the time we talked.)

The whole-genome approach had also yielded some surprises for Cole. Two of my genes changed dramatically over time, one toward way more expression and one toward way less. Cole admittedly didn't know a whole lot about those particular genes, so we looked them up. One was called the BEST1 gene, which had increased its RNA presence sevenfold between Time One and Time Three. When, together over our video call, we consulted the wiki of genes, BEST1, short for bestrophin-1, seemed to be well named. It's a good gene to be turning on, because it protects eyes and facilitates the transport of glucose. That was something I needed.

At the same time, expression in my so-called AGER gene had dropped 90 percent. More good news. AGER, as might be expected, is a gene associated with aging, including trouble with glucose biology, Alzheimer's, and diabetes-related blood-vessel disease.

"You couldn't come up with a more charming set of results," said Cole. "You do not look like a chronically lonely person. You may have your ups and downs, but somehow your brain stem is sitting there saying, it's basically okay, I'm not fundamentally alone in the world. I'm not bereft of support or options."

I would have uncorked champagne right there except it might have sent AGER on a reverse course, and I wanted to stay at my BEST. Cole continued on in a comforting way, engaging his Jungian mode.

"The other thing that's important to know about you is that you have a really big kind of North Star," he said. Like me, he believes in journalism, and in its social value. "You've got a thing that you're doing, you've got a life mission, and that is the most reliable kind of psychology we've seen for bumping aside threat. In the end, the more I talk to you," he continued, "the more I think that that does seem like your dominant sort of psychobiology. You have your challenges and disappointments and your discoveries and disillusions, but you've got a really strong sense of who you are and what you're about and why you're doing what you're doing. And that doesn't move a centimeter across all of the months that I've talked to you. In terms of the personality of your body, I think that's probably the defining thing."

Finally, after all these many months, I truly believed he was right. If Steve Cole ever gets tired of transcription factors, he's got a future in divorce counseling. Happily, he also has a future in personalized medicine. We seemed to know what was working for me—social support, time in nature (note to self: sometimes bring friends and family), and purpose—and those things undoubtedly work for many people. But wouldn't it be nice if everyone could take their before-and-after blood mood after every large and small behavioral adjustment and know for sure if it's helping them as they get buffeted by the vicissitudes of love and life?

That is exactly the future that Cole wants to build. "There's no one right answer for everybody. Different things work for different people. The best thing is to actually measure your own biology while you try menu item A like meditating for two months. Maybe it will change your leukocytes forever. Maybe it won't." We are not always good judges of what helps us, he pointed out. What he wants is to shrink the cost and time required for personalized RNA data so that everyone can have real-time molecular biofeedback.

Such a future may be a ways off. In the meantime, there are some other indicators of how we're doing. Do we feel safe? Does our world make sense to us or is it chaotic and unpredictable? How well are we sleeping? We can't always read our leukocytes, but we can read our moods. Are we treating people with patience? The Buddhists say enlightenment looks like love: seeing the good in everyone from a place of openness, calm, and compassion. If you're able to do that, your transcription factors are probably as shiny as platinum. Keep doing the things you love, and then try, per Cole, some things you might not, things that take effort. Don't forget the less obvious, evidence-based suggestions for flourishing in this crazy world: purpose, connection, biophilia. And, of course, Cole's favorite for healing heartbreak and its attendant afflictions: "Focus outward on other people or the world or history. Make a contribution."

These were good lessons for the pandemic as well, now that the themes I explored in heartbreak were cresting all over again: shock, isolation, vulnerability, grief. This time the experience was undeniably shared. Heartbreak had become larger than my story, as heartbreak always was and as it is meant to be. Now, though, I knew I was better fortified. The insights I'd gleaned from the loss of love—and with it, the shaky expectations of safety and a sense of the future—felt more important than ever.

My leukocytes weren't yet prismatic, light-filled precious gems, but they were no longer brittle bits of coal.

Cole's final words of advice to me?

"Keep doing what you're doing. The trajectories look good."

25

A BOAT OF LETTUCE

When it comes to learning, Triumph is the real foe;
it's Disaster that's your teacher.
—MARIA KONNIKOVA, *THE BIGGEST BLUFF*

This book starts with a river and it ends with a river.

I hadn't brought an object with me to the Museum of Broken Relationships. By the time I went, I'd jettisoned many artifacts of the marriage, but I still had a big one: my wedding ring. It sat in a cardboard jewelry tray my daughter made in sixth grade. I'd come to admire the idea of ritual purgation. The Japanese, who love rituals and who also divorce at high rates, have a clever solution that is growing increasingly popular. A couple, or sometimes one-half of a couple, will hire a divorce-ceremony planner to help perform it. The planner will create everything you need for a proper divorce celebration: food, flowers, and a large mallet with which to smash your wedding ring. You can hold the mallet by yourself or with your ex.

While pounding the ring does sound satisfying, it wasn't really the sentiment I was going for. A friend told me about a woman who pawned her ring after her breakup. She received quite a lot of cash for it, which she handed over to the first homeless person she saw. I loved that, but my ring was pretty chintzy—we used to joke it was like the hoop on a soda top—and because we were still in pandemic lockdown in DC in early May, pawnshops were closed. Where I really wanted to go, not surprisingly, was the river.

Three years to the day after my ex moved out, I fish the ring out of the cardboard. I'd safeguarded this slender hoop since I was 25. I invite Lauri and Eliza, my friends who often walk with me, to join the hike down to the banks of the Potomac. They immediately sign on. The river is surging after long days of April rain. A white sedan has careened off the towpath and cowers hoodfirst like an embarrassed swan in the middle of the canal that parallels the river. It seems emblematic of our quarantines—stuck in place, journey interrupted—and also a strange and somehow apt incursion of the urban into the wild.

I considered attaching the ring to a rock and hurling it into the gurgling river or dropping it from a great height off Chain Bridge. Both of those seemed too aggressive as well as metaphorically inept. You can't just sink a marriage under the water and expect it not to resurface. That much I had learned. I had been undertaking rituals all along, albeit ones largely rooted in science. I had tried writing narratives, indulging in negative appraisals, finding solace and distraction in heat, drugs, and other temptations, burning missives, cursing, seeking awe, walking it off. Even sacrificing my blood to UCLA felt ritualistic, an arterial offering in a bargain for insight and knowledge. I'd hoped, through this long process, to find the terminus of pain. But resolution doesn't come that easily. Most of the things I'd tried had helped, some hadn't. The best I could hope for now was distance, perspective, and the passage of time. A fast-flowing river, I knew, was the right place for that.

The Potomac isn't as glamorous or as mighty as the Green, but it's my home river now, and I've gotten to like it. I know its birds and some of its fishes. I know its floods and its shores and its reeds and the hidden trails. You can see two states plus the District from one spot. George Washington lived in Mt. Vernon along its banks. John Brown attempted to create a stronghold nearby as a stand against slavery. General Robert E. Lee crossed it twice to invade the North. In its flood-plain, my friends and neighbors, inveterate scavengers, have found squat glass decanters from the 1940s, lead bullets from the Civil War, brass Victorian doorknobs, arrowheads, and a message in a bottle that was launched 40 canal locks away in 1975 and that still smells like brandy. Walking along this river, the fourth largest on the Atlantic coast, has been my salvation, my near-daily infusion of beauty and grace. I often bring my kids, and we picnic and walk. During the lockdown, sometimes I come alone, sometimes in a face mask, feeling like Butch Cassidy waiting for his escape pony.

I no longer dread being alone; in fact, I treasure time by myself. I no longer need to pinwheel about, too anxious or frenetic to stay still. I've grown more comfortable with uncertainty. I'm better at finding beauty in the everyday and I now look for it, often. I'm more aware of the suffering of others, and more determined to help alleviate it. I know I have plenty of love to give.

What I'm not is someone with no regrets and an easy, nourishing relationship to my past. I'm still not great friends with my ex, although we get along. He has, on several occasions, sincerely apologized for the pain he caused in the breakup. I still have dreams about him, especially around significant dates like these anniversaries. In the dreams, we're often on trails or rivers, with friends and our kids, traveling together in a wondrous landscape. I wake up missing that shared adventure, still. I still wonder, sometimes, why it disappeared and then I feel briefly disoriented. I don't expect to lose all these emotions, nor do I need to lose them altogether.

Finally, too, I have come to feel some measure of forgiveness. It didn't appear because he asked for it or because I ingested more psychedelics. I'm not sure why it suddenly showed up, but it's probably attributable to the best heartbreak cure of all: time.

A word on time. I believe all of the things I tried probably sped up my heartbreak convalescence between 25 and 50 percent. The literature predicts it takes three to four years for one's emotional and physical health to return to baseline, at least after a long marriage. I didn't have pre-breakup blood to compare, but at two years out from our split, my biomarkers were getting sparkly and at three years out, I know I am feeling like a fuller, keener, softer, wiser version of myself than ever before.

So instead of a mallet, I choose lettuce. Like a heart, it's supple and impermanent. I've threaded the ring onto a thin wooden skewer, and then run the skewer through a crisp romaine leaf. I crisscross a couple of other skewers across it, making a floating nest. It is a hot and sunny day after all that rain. The wrens and the red-winged blackbirds call overhead. Mallards lark about in the eddies, two by two. I'm glad, truly, that pair-bonding is working for somebody. Also, I like remembering that fish eggs still hatch after being eaten and excreted by ducks.

I still believe in love. I still believe that for many of us, a strong partnership can be a wonderful way to move through an unpredictable world. In its absence, there are alternatives. The best are family, community, and an inkling of optimism that while, yes, the world may sometimes seem chaotic and unjust, it is also beautiful and ever changing. Somehow most things will work out in the end if we can just help them along.

By the time we find a serviceable beach where the current moves in close, the lettuce boat is wilting. We must hurry before it becomes a sad salad. My boots aren't quite high enough, so Eliza gives me hers. The river buckles and pulls against my legs as it charges its way past Virginia, then Maryland, to the Chesapeake Bay on its eventual ride to the Atlantic Ocean. I wade out, finding my balance while cupping the boat in both hands. The ring, strapped on like Major Powell, sparkles in its perch. I nod in appreciation and release it to the rush of waters.

Acknowledgments

I t takes a lot of support to heal a heart, and also to write a book.

I owe a huge debt to the many scientists, friends, and colleagues who let me blather on about both endeavors and who kept me going. First, to the scientists who welcomed me into their labs, offices, and even homes: Steve Cole, Paula Williams, Kimberley Johnson, Ty McKinney, Zoe Donaldson, Tor Wager, Moriel Zelikowsky, Jay Love, Naomi Eisenberger, Lani Shiota, Ryan Hampton, Dacher Keltner, Craig Anderson, Bert Uchino, David Strayer, Rob Kent de Grey, Richard Smeyne, Helen Fisher, Andrew Steptoe, Nathan DeWall, Sian Harding, and Liam Couch.

Thanks also to the practitioners and general movers and shakers who shared their expertise: Chelsea Van Essen, Elise Knicely, Aleya Littleton, Stacy Bare, Kelly Smyth-Dent, Denise Mitten, Mims Davies, Roland Duke, Killian Noe, Traci Sooter, Chris Cotten, Amy Chan, Nancy Wiens, Julie Barnes, Caron Curragh, Natalie Goldberg, and Lisa Herrick.

For river advice, support, and logistics: my father, John Williams, champion rower Berkeley Williams, Doug Dupin, Allen O'Bannon, Bob Ratcliffe, Mike Fiebig, Emily Scott, Sara LoTemplio, Steve Markle and the OARS team, Tim Palmer and Ann Vileisis, Doug Robotham, Lise

Aagenbrug, Violet Wallach, Kevin Lake, Deb Love, Matt Madden, and Tim Sullivan. I would paddle with you anywhere.

Many friends and relatives provided housing and succor during reporting or writing: Mara Rabin and Kevin Shilling, Peter Williams and Lisa Jones, Meg Dawson, Gary Nabhan and Laurie Monti, Sam Lawson and Laurel Mayer, Kelly Cash, Dave and Rebecca Livermore, Graham Chisholm, Shanti and Mark Hodges, Chuck Slaughter, Molly West, Ann Skartvedt, Maude Meyers, Lucy Hall and Joe Smalley, Kirkland Newman Smulders, Buzzy Jackson, Kita and Toby Murdock, Auden Schendler and Ellen Friedman, and Pamela Geismar and Pete Friedrich. Sometimes (okay, often), a writer needs to put her head down without looking up. I spent a productive weeklong residency in the Georgia pines courtesy of Steve Nygren and AIR Serenbe's artists' cabin, and another one at the Virginia Center for the Creative Arts just before the pandemic sent us all back home.

I'm grateful to Melissa Perry and the Department of Environmental and Occupational Health at George Washington University for sponsoring my professorial lectureship, which comes with excellent library access and a collegial community. Over the years of writing this book, I received helpful research assistance from Emily Ounanian, Mara Abbott, and Amanda Eggert. Even though a pandemic kept me from Japan, I appreciate the legwork by Atsuko Horiguchi and Junko Taniguchi.

Thanks to Mary Beth Kirchner, Martha Little, and the good people at Audible, which commissioned our audio production *The 3-Day Effect*. Some of the interviews for that informed ideas loosely reflected in this book. Also thanks to *Outside* magazine, especially Mary Turner, Mike Roberts, and Peter Frick-Wright, for assigning and shaping the trafficking survivors' story for both print and audio.

Hannah Nordhaus is one of those uncommonly talented people who is both an exquisite writer and a brilliant editor. I owe her a big read or three. Elizabeth Hightower Allen jumped right from editing my magazine articles at *Outside* into editing my book. She's the best. Tom Hjelm provided sage advice at many turns, as well as emotional support and a steady (modest) supply of Kentucky straight. Others lent wise counsel and editorial suggestions on portions of the manuscript or the whole enchilada: Naomi Williams, Lisa Jones, Hillary Rosner, Jay Heinrichs, Margaret Nomentana, Eliza McGraw, Lauri Menditto, David Plotz, Josh Horwitz, Erica Perl, Jacki Lyden, and Eric Weiner. Thanks also to Howard Norman, Linda Reveal, Ann Mah, David Grinspoon, Juliet Eilperin, Matt Davis, David Ginosar, Kirk Johnson, and Ruth Lichtman. I'm grateful to my ex-husband for understanding why I needed to write this book.

Immense gratitude goes to my bighearted agent Molly Friedrich, and to Lucy Carson. I've been fortunate to work with Jill Bialosky as my editor for three books, and I'm so grateful for her prodigious faith, skill, and guidance. Thanks also to the rest of the wonderful team at W. W. Norton, including Erin Lovett, Steve Colca, Meredith McGinnis, and Drew Weitman. Many thanks for excellent copyediting by Sarah Johnson.

I couldn't have made it through these difficult years without the unwavering love and support of so many friends and relatives. Even though I may not want them to read this book for a while, Ben Williams and Annabel Williams buoyed me, sheltered me, and inspired me. My heart will always belong to them.

Notes

INTRODUCTION

1   some plastic bags: Let it be noted for the record that I do not advocate violating federal regulations on the toilet system. If you are running the river solo, as of this writing you are allowed to use so-called Wag Bags, which are basic sealable plastic bags with a few accoutrements. I did have plenty of these as a backup on the trip, and they are the bomb.

5   About 39 percent of all first marriages in the US end in divorce: Belinda Luscombe, "The Divorce Rate Is Dropping. That May Not Actually Be Good News," *Time*, November 26, 2018.

5   1852 medical textbook: G. B. Wood, *A Treatise on the Practice of Medicine* (Philadelphia: Lippincott, 1852).

5   Divorce as a top stressful life event: See the Life Changes Stress Test, http://www.dartmouth.edu/eap/library/lifechangestresstest.pdf, based on T. H. Holmes and R. H. Rahe, "The Social Readjustment Rating Scale," *Journal of Psychosomatic Research* 11, no. 213 (1967); and B. L. Bloom, S. J. Asher, and S. W. White, "Marital Disruption as a Stressor: A Review and Analysis," *Psychology Bulletin* 85 (1978): 867–94.

5   Catullus's word is part of poem 85, *Odi et amo*, "I Love and I Hate."

5   Regarding Sontag and Jasper Johns, see Benjamin Moser, "Open to Interpretation: The Brief Relationship of Susan Sontag and Jasper Johns," Literary Hub, September 19, 2019.

6   weep-dancing while belting out Gloria Gaynor: I'd just like to say, the Cake version of "I Will Survive" is truly excellent. I recommend it.

CHAPTER 1: BRIDGE TO NOWHERE

13   Creaking Bridge study: D. G. Dutton and A. P. Aron, "Some Evidence for Heightened Sexual Attraction under Conditions of High Anxiety," *Journal of Personality and Social Psychology* 30, no. 4 (1974): 510–17.

15   Notlim: Milton and Deirdre are not their real names.

CHAPTER 2: THE HEART

24   American Heart Association recognized Takotsubo in 2006: Scott W. Sharkey and Barry J. Maronn, "Epidemiology and Clinical Profile of Takotsubo Cardiomyopathy," *Circulation Journal* 78, no. 9 (2014): 2119.

24   patients appear to be having a regular heart attack: Jessica Ebert, "A Broken Heart Harms Your Health," *Nature*, January 9, 2005, https://www.nature.com/news/2005/050207/full/050207-11.html.

24   a portion of the left ventricle: M. Sato et al., "Increased Incidence of Transient Left Ventricular Apical Ballooning (So-Called 'Takotsubo' Cardiomyopathy) after the Mid-Niigata Prefecture Earthquake," *Circulation Journal* 70, no. 8 (2006): 949.

24   Risk of complications after Takotsubo: Anthony Kashou et al., "A Case of Recurrent Takotsubo Cardiomyopathy: Not a Benign Entity," *Journal of Medical Cases* 9, no. 3 (2018): 98.

25   The Niigata Prefecture quake: see M. Sato et al. The Ohio study during the pandemic: Jabri A, Kalra A, Kumar A, et al., "Incidence of Stress Cardiomyopathy During the Coronavirus Disease 2019 Pandemic," *JAMA Network Open* 3, vol. 7 (2020).

25   For accounts of the spectator deaths, see: S. Y-Hassan, K. Feldt, and M. Stålberg, "A Missed Penalty Kick Triggered Coronary Death in the Husband and Broken Heart Syndrome in the Wife," *American Journal of Cardiology* 116, no. 10 (2015): 1639–42; M. Fijalkowski et al., "Takotsubo Cardiomyopathy in a Male during a Euro 2012 Football Match," *Clinical Research in Cardiology* 102, no. 4 (2013): 319–21.

25   A survey of 43 million medical records in Denmark found that in the year following a romantic breakup: Margit Kriegbaum et al., "Does the Association between Broken Partnership and First Time Myocardial Infarction Vary with Time after Break-Up?," *International Journal of Epidemiology* 42 (2013): 1816.

26   As physician Sir William Osler put it in 1908: "The tragedies of life are largely arterial." For this quote, from *Diseases of the Circulatory System*, I am indebted to Sandeep Jauhar, *Heart: A History* (New York: Farrar, Straus and Giroux, 2018).

26   80 percent of Takotsubo cases occur in postmenopausal women: Christian Templin et al., "Takotsubo Syndrome: Underdiagnosed, Underestimated, but Understood?," *Journal of the American College of Cardiology* 67, no. 16 (2016): 1937.

26   For accounts of the Joplin tornado, I'm relying on my interviewing notes and also excellent reporting by the *Kansas City Star*, such as: Cindy Hoedel and Lisa Gutierrez, "The Gathering Storm: Tracing the Trail of Joplin's Killer Tornado," *Kansas City Star*, December 9, 2011, https://www.kansascity.com/news/special-reports/article299994/The-gathering-storm-Tracing-the-trail-of-Joplin's-killer-tornado.html.

29   30 percent of people might show symptoms of PTSD: George A. Bonanno et al., "Weighing the Costs of Disaster: Consequences, Risks, and Resilience in Individuals, Families, and Communities," *Psychological Science in the Public Interest* 11, no. 1 (2010): 3.

CHAPTER 3: HINDU KUSH

34   one of the few researchers to study the brains of people who have been dumped: Helen E. Fisher et al., "Reward, Addiction, and Emotion Regulation Systems Associated with Rejection in Love," *Journal of Neurophysiology* 104 (2010): 51–60.

37   Suicide attempts in adolescents: E. Paul, "Proximally-Occurring Life Events and the First Transition from Suicidal Ideation to Suicide Attempt in Adolescents," *Journal of Affective Disorders* 241 (2018): 499–504.

37   Suicide in adults: T. Chen and K. Roberts, "Negative Life Events and Suicide in the National Violent Death Reporting System," *Archives of Suicide Research*, published online October 22, 2019: 1–15.

37   Suicide in India: From Larry Young and Brian Alexander, *The Chemistry between Us* (New York: Penguin Random House, 2014), Kindle.

39   women ruminate or think about their love troubles more than men: J. E. Graham et al., "Marriage, Health, and Immune Function," in *DSM-V: Neuroscience, Assessment, Prevention, and Treatment*, ed. Steven R. Beach et al. (Washington, DC: American Psychiatric Association, 2007), 75–93. Also see J. Kiecolt-Glaser and T. Newton, "Marriage and Health: His and Hers," *Psychological Bulletin* 127, no. 4 (2001): 472–503.

CHAPTER 4: A COSTLY LIFE EVENT

42   The Jane Austen quote is from a letter to her 24-year-old niece, Fanny, in 1817. I still take heart in a later bit from the letter: "[Do not] be in a hurry; depend upon it, the right Man will come at last; you will in the course of the next two or three years, meet with somebody more generally unexceptionable than anyone you have yet known, who will love you as warmly as ever He did, & who will so completely attach you, that you will feel you never really loved before." (It was good advice; she ended up marrying a baronet.) Letter of March 13, 1817, found in Vivien Jones, ed., *Jane Austen: Selected Letters* (Oxford: Oxford University Press, 2004), 204.

43   Unhappy zebra fish: cited in India Morrison, "Keep Calm and Cuddle On: Social Touch as a Stress Buffer," *Adaptive Human Behavior and Physiology* 2 (2016): 344–62.

43   Cortisol levels when long-term couples are apart: D. Saxbe and R. L. Repetti, "For Better or Worse? Coregulation of Couples' Cortisol Levels and Mood States," *Journal of Personality and Social Psychology* 98 (2010): 92–103. See also L. M. Diamond, A. M. Hicks, and K. D. Otter-Henderson, "Every Time You Go Away: Changes in Affect, Behavior, and Physiology Associated with Travel-Related Separations from Romantic Partners," *Journal of Personality and Social Psychology* 95, no. 2 (2008): 385–403.

43   Brain wave study in couples: S. Kinreich et al., "Brain-to-Brain Synchrony during Naturalistic Social Interactions," *Scientific Reports* 7 (2017): 17060.

46   Scores of robust studies: David A. Sbarra and Paul J. Nietert, "Divorce and Death: Forty Years of the Charleston Heart Study," *Psychological Science* 20 (2009): 107–13.

46   For a good overview of the health effects of marriage, see the review of the literature by David Sbarra: Matthew E. Dupre, Audrey N. Beck, and Sarah O. Meadows, "Marital Trajectories and Mortality among US Adults," *American Journal of Epidemiology* 170, no. 5 (2009): 546–55.

46   A large Danish study: K. Laugesen et al., "Social Isolation and All-Cause Mortality: A Population-Based Cohort Study in Denmark," *Scientific Reports* 8 (2018): 4731.

46 Quote from William Farr: David A. Sbarra, Rita W. Law, and Robert M. Portley, "Divorce and Death: A Meta-Analysis and Research Agenda for Clinical, Social, and Health Psychology," *Perspectives on Psychological Science* 6, no. 5 (2011): 454.

46 The good-genes theory doesn't explain the full effect of healthier married people: J. E. Graham et al., "Marriage, Health, and Immune Function," in *DSM-V: Neuroscience, Assessment, Prevention, and Treatment*, ed. Steven R. Beach et al. (Washington, DC: American Psychiatric Association, 2007): 75–93.

47 The single largest determinant of health in the US is wealth: Angus Deaton, "Policy Implications of the Gradient of Health and Wealth," *Health Affairs* 21, no. 2 (2002): 13–30. See also Bruce G. Link and Jo Phelan, "Social Conditions as Fundamental Causes of Disease," *Journal of Health and Social Behavior*, special issue (1995): 80–94.

47 Married people are more likely to accumulate resources, afford health insurance, etc.: A good summary of this literature can be found in L. J. Waite and E. L. Lehrer, "The Benefits from Marriage and Religion in the United States: A Comparative Analysis," *Population and Development Review* 29, no. 2 (2003): 255–76.

47 For the health penalty of bad marriages, see M. A. Whisman, A. L. Gilmour, and J. M. Salinger, "Marital Satisfaction and Mortality in the United States Adult Population," *Health Psychology* 37, no. 11 (2018): 1041–44. Also see Julianne Holt-Lunstad, Wendy Birmingham, and Brandon Jones, "Is There Something Unique about Marriage? The Relative Impact of Marital Status, Relationship Quality, and Network Social Support on Ambulatory Blood Pressure and Mental Health," *Annals of Behavioral Medicine* (May 2008).

47 "boundless loneliness": Nicole Krauss, *Forest Dark* (New York: Harper Perennial, 2017), 42.

48 women in same-sex marriages may be happier: Stephanie Coontz, "How to Make Your Marriage Gayer," *New York Times*, February 13, 2020.

48 Divorced men languish: One study found that unmarried women have a 50 percent greater risk of mortality than do similar but married women, compared with a 250 percent greater risk for men. See Catherine Ross, John Mirowsky, and Karen Goldsteen, "The Impact of the Family on Health: The Decade in Review," *Journal of Marriage and the Family* 52 (1990): 1059.

48 new evidence that wealth really is linked to happiness: J. M. Twenge and A. B. Cooper, "The Expanding Class Divide in Happiness in the United States, 1972–2016," *Emotion*, advance online publication (2020).

49 "Marriage is no picnic": Edward B. Foote, *Dr Foote's Home Cyclopedia of Popular Medical, Social and Sexual Science* (New York: Murray Hill Publishing Co., 1900), 1051.

49 A study by the National Institute on Aging in 2000: P. T. Costa Jr. et al., "Personality at Midlife: Stability, Intrinsic Maturation, and Response to Life Events," *Assessment* 7, no. 4 (2000): 365–78.

49 The Charleston study: Sbarra and Nietert, "Divorce and Death: Forty Years."

49 A large analysis from 11 countries: Sbarra, Law, and Portley, "Divorce and Death: A Meta-Analysis," 454–74.

50 Divorce is "a costly life event": Dupre, Beck, and Meadows, "Marital Trajectories and Mortality."

50 Around 15 percent of people don't get over divorce: Sbarra, Law, and Portley, "Divorce and Death: A Meta-Analysis."

52 Openness is characterized as comfort with novelty: Colin G. DeYoung et al., "Openness to Experience, Intellect, and Cognitive Ability," *Journal of Personality Assessment* 96 (2014): 46–52. See also Paula G. Williams et al., "Individual Differences in Aesthetic Engagement Are Reflected in Resting-State fMRI Connectivity: Implications for Stress Resilience," *NeuroImage* 179, no. 1 (2018): 156.

52 William Blake's wife, quoted by Jonathan Bate: Melvyn Bragg and Jonathan Bate, "Songs of Innocence and of Experience," *In Our Time*, BBC4, June 23, 2016; the quote is referring to a passage from Algernon Charles Swinburne, *William Blake: A Critical Essay* (Oxford: John Camden Hotten, 1868), and cited by: Edwin J. Ellis, *The Poetical Works of William Blake* (London: Chatto and Windus, 1906), 257.

52 Beethoven quoted in J. W. N. Sullivan, *Beethoven: His Spiritual Development* (New York: Mentor Books, 1949), 92.

54 Other research supporting Williams's theory: The most interesting are the studies looking at the neural correlates of awe under psilocybin. See Robin L. Carhart-Harris et al., "Neural Correlates of the Psychedelic State as Determined by fMRI Studies with Psilocybin," *Proceedings of the National Academy of Sciences* 100, no. 15 (2012): 8788–92. See also A. V. Lebedev et al., "Finding the Self by Losing the Self: Neural Correlates of Ego-Dissolution under Psilocybin," *Human Brain Mapping* 36, no. 8 (2015): 3137–53.

54 Study showing a "tendency to orient oneself toward a larger transcendent reality": V. Saroglou, C. Buxant, and J. Tilquin, "Positive Emotions as Leading to Religion and Spirituality," *Journal of Positive Psychology* 3 (2008): 165–73.

55 Blake's letter: Bragg and Bate, "Songs of Innocence and of Experience."

CHAPTER 6: ALL PAIN IS ONE MALADY: REJECTION

73 James quoted in Kipling D. Williams, "Ostracism," *Annual Review of Psychology* 58 (2007): 426.

76 Williams wrote that feeling rejected in this way increases blood pressure: See Williams, "Ostracism," 425–52.

76 They performed the worst on a logical-reasoning task and consumed more food: Roy F. Baumeister et al., "Social Exclusion Impairs Self-Regulation," *Journal of Personality and Social Psychology* 88, no. 4 (2005): 589–604.

77 Brain scans showed less activation in executive function: Keith W. Campbell et al., "A Magnetoencephalography Investigation of Neural Correlates for Social Exclusion and Self-Control," *Social Neuroscience* 1, no. 2 (2006): 127.

77 "loss of self-control": Baumeister, "Social Exclusion Impairs Self-Regulation," 592.

77 ancient Greeks: Described in Williams, "Ostracism," 428.

77 Ostracism in animals: Marc Bekoff and Jessica Pierce, *Wild Justice: The Moral Lives of Animals* (Chicago: University of Chicago Press, 2009), 1–12.

78 In a rather elaborate experiment at Northeastern University: David DeSteno, Piercarlo Valdesolo, and Monica Bartlett, "Jealousy and the Threatened Self: Getting to the Heart of the Green-Eyed Monster," *Journal of Personality and Social Psychology* 91, no. 4 (2006): 626–41.

78 Men are far more violent in relationship conflict: E. Petrosky et al., "Racial and Ethnic Differences in Homicides of Adult Women and the Role of Intimate Partner Violence—United States, 2003–2014," *Morbidity and Mortality Weekly Report* 66, no. 28 (July 21, 2017): 741–46.

80 Wager's team was one of the first to try to quantify the overlap between the physical and social pain centers: Ethan Kross et al., "Social Rejection Shares Somatosensory Representations with Physical Pain," *PNAS* 108, no. 15 (2011): 6270–75.

81 Placebo subjects showed less intense heartache: Leonie Koban et al., "Frontal-Brainstem Pathways Mediating Placebo Effects on Social Rejection," *Journal of Neuroscience* 37, no. 13 (2017): 3621–31.

CHAPTER 7: HEARTBREAK HOTEL: GRIEF

83 Monogamy in less than 10 percent of mammals: Jeffrey A. French et al., "Social Monogamy in Nonhuman Primates: Phylogeny, Phenotype, and Physiology," *Journal of Sex Research* (2017): 1–25; and Dieter Lukas and Timothy Hugh Clutton-Brock, "The Evolution of Social Monogamy in Mammals," *Science* 341, no. 6145 (2013): 526–30.

83 finding a long-term partner increases the brain's metabolism of fuel in the form of glucose: N. Maninger et al., "Pair Bond Formation Leads to a Sustained Increase in Global Cerebral Glucose Metabolism in Monogamous Male Titi Monkeys (Callicebus Cupreus)," *Neuroscience* 348 (2017): 302–12.

83 Termites: Henry Gee, "Domestic Violence and Divorce, Termite-Style," *Nature*, published online January 28, 1999.

87 75 percent of prairie vole couples stay together: Lowell L. Getz and C. Sue Carter, "Prairie-Vole Partnerships," *American Scientist* 84 (1996): 56.

87 20 percent of males will pair up with someone new: C. Sue Carter, A. Courtney DeVries, and Lowell L. Getz, "Physiological Substrates of Mammalian Monogamy: The Prairie Vole Model," *Neuroscience and Biobehavioral Reviews* 19, no. 2 (1995): 303–14.

87 They even console each other: Tobias T. Pohl, Larry J. Young, and Oliver J. Bosch, "Lost Connections: Oxytocin and the Neural, Physiological, and Behavioral Consequences of Disrupted Relationships," *International Journal of Psychophysiology* 136 (2019): 58.

89 the nucleus accumbens is unusually active while looking at pictures of lost loved ones: Harald Gündel et al., "Functional Neuroanatomy of Grief: An FMRI Study," *American Journal of Psychiatry* 160, no. 11 (2003): 1946–53.

89 Donaldson can make vole neurons fire: The fascinating process of optogenetics is nicely described in Karl Deisseroth, "Optogenetics: Controlling the Environment with Light [Extended Version]," *Scientific American*, October 20, 2010.

90   Bosch's study: Oliver J. Bosch et al., "The CRF System Mediates Increased Passive
      Stress-Coping Behavior Following the Loss of a Bonded Partner in a Monogamous
      Rodent," *Neuropsychopharmacology* 34, no. 6 (2009): 1406–15.

91   The despondent singletons spent more time in dark boxes than exploring other rooms:
      P. Sun et al., "Breaking Bonds in Male Prairie Vole: Long-Term Effects on Emotional
      and Social Behavior, Physiology, and Neurochemistry," *Behavioral Brain Research* 265
      (2014): 22–31.

CHAPTER 8: WELCOME TO THE EREMOCENE: ATTACHMENT

95   The divorce rate I cite refers to a figure of 14.4 percent for women with post-college
      degrees after 10 years of marriage: Steven P. Martin, "Trends in Marital Dissolution by
      Women's Education in the United States," *Demographic Research* 15 (2006): 537–60.

95   Divorce rate has fallen by nearly half: Martin, "Trends in Marital Dissolution."

97   On the history of the word *anxiety* see: Marc-Antoine Crocq, "A History of Anxiety:
      From Hippocrates to DSM," *Dialogues in Clinical Neuroscience* 17, no. 3 (2015): 319.

98   The cursing study: Richard Stephens, John Atkins, and Andrew Kingston, "Swearing
      as a Response to Pain," *Neuroreport* 20 (2009): 1056–60.

99   Laing quote from Olivia Laing, *The Lonely City: Adventures in the Art of Being Alone*
      (New York: Macmillan, 2016), 13.

100   Watson quoted in Deborah Blum, *Love at Goon Park* (New York: Basic Books, 2011), 37,
      from John B. Watson and Rosalie Alberta Watson, "The Dangers of Too Much Mother
      Love," in *Psychological Care of Infant and Child* (New York: W. W. Norton, 1928).

100   Freud's quote on parental love is paraphrased here but cited in full: Watson and Wat-
      son, "The Dangers of Too Much Mother Love," 7.

100   many Western hospitals separated mothers and infants: Rima D. Apple, *Mothers and
      Medicine: A Social History of Infant Feeding, 1890–1950* (Madison: University of Wis-
      consin Press, 1987). See also Margo Edwards and Mary Waldorf, *Reclaiming Birth:
      History and Heroines of American Childbirth Reform* (Trumansburg, NY: Crossing
      Press, 1984).

100   "Nothing in psychology had predicted this": Blum, *Love at Goon Park*, 46.

101   Bowlby human spirit quote: From a lecture originally published in 1968 in John
      Bowlby, *The Making and Breaking of Affectional Bonds* (London: Tavistock Publica-
      tions, 1979), 66.

101   The wire mother experiments are described in Blum, *Love at Goon Park*, 150.

102   The isolation experiments: Blum, 213–15.

102   Depression quote: Blum, 215.

102   Dogs helped the lonely, scarred monkeys: Jaak Panksepp, *Affective Neuroscience: The
      Foundations of Human and Animal Emotions* (Oxford: Oxford University Press,
      1998), 275.

105   Socially isolated mice attack submissive intruders: M. Zelikowsky et al., "The Neuro-
      peptide Tac2 Controls a Distributed Brain State Induced by Chronic Social Isolation
      Stress," *Cell* 173, no. 5 (2018): 1269.

CHAPTER 9: YOUR CELLS ARE LISTENING

108 For more on the inflammation-disease link, see J. Morey et al., "Current Directions in Stress and Human Immune Function," *Current Opinion in Psychology* 5 (2015): 13–17.

108 High-pathogen areas and risks from inflammation: Andrew H. Miller and Charles L. Raison, "The Role of Inflammation in Depression: From Evolutionary Imperative to Modern Treatment Target," *Nature Reviews Immunology* 16 (2016): 22.

109 For more about Michael Snyder's personalized medicine experiments, see Xiao Li et al., "Digital Health: Tracking Physiomes and Activity Using Wearable Biosensors Reveals Useful Health-Related Information," *PLoS Biology* 15 (2017): 1–30.

109 Researchers at Ohio State: Janice K. Kiecolt-Glaser et al., "Marital Quality, Marital Disruption, and Immune Function," *Psychosomatic Medicine* 49 (1987): 3–34.

109 Sbarra's review paper: David A. Sbarra, Rita W. Law, and Robert M. Portley, "Divorce and Death: A Meta-Analysis and Research Agenda for Clinical, Social, and Health Psychology," *Perspectives on Psychological Science* 6, no. 5 (2011): 454–74.

110 looked at a group of 72 men: Steve W. Cole, Margaret E. Kemeny, and Shelley E. Taylor, "Social Identity and Physical Health: Accelerated HIV Progression in Rejection-Sensitive Gay Men," *Journal of Personality and Social Psychology* 72, no. 2 (1997): 320–35. Also see: Steve W. Cole et al., "Elevated Physical Health Risk among Gay Men Who Conceal Their Homosexuality," *Health Psychology* 15 (1996): 243–51.

111 Chronic loneliness increases the risk of early death by 26 percent: Julianne Holt-Lunstad, Timothy B. Smith, and J. Bradley Layton, "Social Relationships and Mortality Risk: A Meta-Analytic Review," *PLoS Medicine* 7, no. 7 (2010): e100316.

111 The first paper to consider the effect of social factors: S. W. Cole et al., "Social Regulation of Gene Expression in Human Leukocytes," *Genome Biology* 8, no. 9 (2007): R189.

112 Cacioppo's "One of the secrets to a good relationship" quote is from Stephen Heyman, "Don't Know What the Angular Gyrus Is? Your Heart Does," *New York Times*, November 18, 2017.

CHAPTER 10: THE BODY DOESN'T LIE

Portions of this chapter appeared in Florence Williams, "The Adventure Therapy Cure for Survivors," *Outside* magazine, May 1, 2018, and in the Outside Podcast.

115 Maner and Miller on how men and women differ: S. L. Miller and J. K. Maner, "Coping with Romantic Betrayal: Sex Differences in Responses to Partner Infidelity," *Evolutionary Psychology* 6, no. 3 (2008): 413–26.

116 One study found that wives are six times more likely to experience major depression: Annmarie Cano and K. Daniel O'Leary, "Infidelity and Separations Precipitate Major Depressive Episodes and Symptoms of Nonspecific Depression and Anxiety," *Journal of Consulting and Clinical Psychology* 68, no. 5 (2000): 774–81.

116 Another study found that, compared to men, women report more self-destructive behaviors: M. R. Shrout and D. J. Weigel, "Infidelity's Aftermath: Appraisals, Mental Health, and Health-Compromising Behaviors Following a Partner's Infidelity," *Journal of Social and Personal Relationships* 35, no. 8 (2018): 1067–91.

116   A number of studies have shown that in heterosexual relationships men get more upset if their mates betray them sexually, while women are more distressed by emotional betrayals: David M. Buss et al., "Sex Differences in Jealousy: Evolution, Physiology, and Psychology," *Psychological Science* 3, no. 4 (1992): 251–55; J. E. Edlund and B. J. Sagarin, "Sex Differences in Jealousy: A 25-Year Retrospective," *Advances in Experimental Social Psychology* 55 (2017): 259–302; B. J. Sagarin et al., "Sex Differences in Jealousy: A Meta-Analytic Examination," *Evolution and Human Behavior* 33 (2012): 595–614.

116   If the betrayed have a so-called "secure" style of relating: Kenneth N. Levy and Kristen M. Kelly, "Sex Differences in Jealousy: A Contribution from Attachment Theory," *Psychological Science* 21, no. 2 (2010): 168–73.

116   A 2017 study at Cornell University: S. Deri and E. M. Zitek, "Did You Reject Me for Someone Else? Rejections That Are Comparative Feel Worse," *Personality and Social Psychology Bulletin* 43, no. 12 (2017): 1675–85.

117   Esther Perel quote: *TED Radio Hour*, NPR, September 11, 2020, https://www.npr .org/2020/09/10/911392320/esther-perel-building-resilient-relationships.

117   In a Gallup poll of over 1,000 Americans: Jeffrey M. Jones, "Most Americans Not Willing to Forgive Unfaithful Spouse," Gallup News, March 25, 2008.

117   Partners stray in an estimated 20–40 percent of heterosexual marriages: Rebeca A. Marín, Andrew Christensen, and David C. Atkins, "Infidelity and Behavioral Couple Therapy: Relationship Outcomes over 5 Years Following Therapy," *Couple and Family Psychology: Research and Practice* 3, no. 1 (2014): 1–12.

117   In about a quarter of divorces, infidelity is cited as a major reason for the split: A. Heintzelman et al., "Recovery from Infidelity: Differentiation of Self, Trauma, Forgiveness, and Posttraumatic Growth among Couples in Continuing Relationships," *Couple and Family Psychology: Research and Practice* 3, no. 1 (2014): 13–29.

123   Freud's view of sexual abuse: Described in Bessel van der Kolk, *The Body Keeps the Score* (New York: Penguin, 2015), 183.

124   Traumatized brains process inputs differently: In van der Kolk, *The Body Keeps the Score*, 93, 249.

CHAPTER 11: SHAGGY BIRDS

133   The Sbarra study on narrative: Kyle J. Bourassa et al., "Tell Me a Story: The Creation of Narrative as a Mechanism of Psychological Recovery Following Marital Separation," *Journal of Social and Clinical Psychology* 36, no. 5 (2017): 359–79.

134   For more on the neurological benefits of exercise, see: Shane O'Mara, *In Praise of Walking* (New York: W. W. Norton, 2019), 137, 155.

138   Emerson's quote, "In the woods, we return to reason and faith . . . ," is from his essay *Nature* (Boston: James Munroe and Company, 1836), 12–13.

138   Stegner's quote is from Wallace Stegner, *The American West as a Living Space* (Ann Arbor: University of Michigan Press, 1988), 15.

139   Kim's paper on rejection and creativity: S. H. Kim, "Outside Advantage: Can Social

Rejection Fuel Creative Thought?," *Journal of Experimental Psychology* 142 (2013): 605–11. See also M. J. C. Forgeard, "Perceiving Benefits after Adversity: The Relationship between Self-Reported Posttraumatic Growth and Creativity," *Psychology of Aesthetics, Creativity, and the Arts* 7, no. 3 (2013): 245–64.

CHAPTER 12: THE WIZARDS OF LONESOME

145 "Western societies have demoted human gregariousness from a necessity to an incidental": John T. Cacioppo and William Patrick, *Loneliness: Human Nature and the Need for Social Connection* (New York: W. W. Norton, 2008), Kindle.

146 The percentage of people living alone in the world is tabulated regularly by Immigroup: Amy Brannan, "Top 10 Loneliest Countries in the World," Immigroup, September 4, 2019, https://www.immigroup.com/news/top-10-loneliest -countries-world. For tabulations by age, see: "Percentage of Americans Living Alone, by Age, 1900–2016," Our World in Data, https://ourworldindata.org/ grapher/percentage-of-americans-living-alone-by-age.

148 Several months into the coronavirus epidemic, loneliness rates had risen: Julianne Holt-Lunstad, Brigham Young University, speaking on a webinar hosted by The Unloneliness Project, May 14, 2020, https://www.hksinc.com/our-news/articles/ connecting-in-strange-times-the-antidote-to-loneliness/.

148 "Let's Talk Loneliness" campaign: UK Office for Civil Society, " 'Let's Talk Loneliness' Campaign Launched to Tackle Stigma of Feeling Alone," press release, June 17, 2019, https://www.gov.uk/government/news/lets-talk-loneliness-campaign-launched -to-tackle-stigma-of-feeling-alone.

CHAPTER 13: TRUTH SERUM, PART ONE

159 Alain de Botton quote from *Essays in Love* (London: Picador, 2006), 199.

159 Bessel van der Kolk quote from his *The Body Keeps the Score* (New York: Penguin, 2014), 263.

160 Three 90-minute sessions: Gail Ironson et al., "Comparison of Two Treatments for Traumatic Stress: A Community-Based Study of EMDR and Prolonged Exposure," *Journal of Clinical Psychology* 58 (2002): 114.

160 In another randomized study of 88 patients with PTSD: Bessel A. van der Kolk et al., "A Randomized Clinical Trial of Eye Movement Desensitization and Reprocessing (EMDR), Fluoxetine, and Pill Placebo in the Treatment of Posttraumatic Stress Disorder: Treatment Effects and Long-Term Maintenance," *Journal of Clinical Psychiatry* 68 (2007): 37.

CHAPTER 14: HIGH ISLAND: WARMTH

164 The negative appraisal study: Christopher Fagundes, "Implicit Negative Evaluations about Ex-Partner Predicts Break-Up Adjustment: The Brighter Side of Dark Cognitions," *Cognition and Emotion* 25 (2011): 164–73.

165 Hornstein's and others' experiments are described in this review paper: Erica A.

Hornstein and Naomi I. Eisenberger, "A Social Safety Net: Developing a Model of Social-Support Figures as Prepared Safety Stimuli," *Current Directions in Psychological Science* 27 (2018): 25–31. Particularly relevant is Naomi I. Eisenberger et al., "Attachment Figures Activate a Safety Signal-Related Neural Region and Reduce Pain Experience," *Proceedings of the National Academy of Sciences* 108, no. 28 (2011): 11721–26.

168   Divorced men and testosterone: Allan Mazur and Joel Michalek, "Marriage, Divorce, and Male Testosterone," *Social Forces* 77 (1998): 315–30.

168   Women who have more sex may go through menopause later: M. Arnot and R. Mace, "Sexual Frequency Is Associated with Age of Natural Menopause: Results from the Study of Women's Health across the Nation," *Royal Society Open Science* 7, no. 1 (2020).

169   I guarantee the Galway Kinnell poem is the raciest poetry you've ever read about a pig: Galway Kinnel, "Saint Francis and the Sow," available online through the Poetry Foundation, https://www.poetryfoundation.org/poems/42683/saint-francis-and-the -sow, and in his *Three Books* (Boston: Houghton Mifflin, 2002).

169   A bunch of papers talk about the opioid network in the context of social pain and attachment: A. Haim, R. J. Van Aarde, and J. D. Skinner, "Burrowing and Huddling in Newborn Porcupine: The Effect on Thermoregulation," *Physiology and Behavior* 52, no. 2 (1992): 247–50; Naomi Eisenberger, "The Pain of Social Disconnection: Examining the Shared Neural Underpinnings of Physical and Social Pain," *Nature Reviews Neuroscience* 13, no. 6 (2012): 421–34; Marianna von Mohr, Louise Kirsch, and Aikaterini Fotopoulou, "The Soothing Function of Touch: Affective Touch Reduces Feelings of Social Exclusion," *Scientific Reports* 7 no. 1 (2017): 1–9; India Morrison, "Keep Calm and Cuddle On: Social Touch as a Stress Buffer," *Adaptive Human Behavior and Physiology* 2 (2016): 344–62.

169   Holding your hand can help mitigate feelings of physical pain: Jarred Younger et al., "Viewing Pictures of a Romantic Partner Reduces Experimental Pain: Involvement of Neural Reward Systems," *PloS One* 5, no. 10 (2010): e13309. See also Sarah L. Master et al., "A Picture's Worth: Partner Photographs Reduce Experimentally Induced Pain," *Psychological Science* 20, no. 11 (2009): 1316–18. See also Pavel Goldstein et al., "Brain-to-Brain Coupling during Handholding is Associated with Pain Reduction," *Proceedings of the National Academy of Sciences* 115, no. 11 (2018): E2528–37. See also Flavia Mancini et al., "Touch Inhibits Subcortical and Cortical Nociceptive Responses," *Pain* 156, no. 10 (2015): 1936–44.

170   Embodied warmth and cold studies are nicely summarized in: John A. Bargh and Idit Shalev, "The Substitutability of Physical and Social Warmth in Daily Life," *Emotion* 12 (2012): 154.

171   Petting dogs: Therese Rehn et al., "Dogs' Endocrine and Behavioural Responses at Reunion Are Affected by How the Human Initiates Contact," *Physiology and Behavior* 124 (2014): 45–53.

171   Oxytocin in rats: studies cited in G. MacDonald and M. R. Leary, "Why Does Social

Exclusion Hurt? The Relationship between Social and Physical Pain," *Psychological Bulletin* 131, no. 2 (2005): 202–23.

171 students in new romantic relationships showed less physical stress: Inna Schneiderman et al., "Love Alters Autonomic Reactivity to Emotions," *Emotion* 11, no. 6 (2011): 1314.

171 Sue Carter quote from the *Relationship School* podcast, no. 320, December 8, 2020.

172 Eagleman quote from David Eagleman, *Livewired: The Inside Story of the Ever-Changing Brain*, illustrated ed. (New York: Pantheon, 2020), 170.

172 Finnish couple study: Ville Renvall et al., "Imaging Real-Time Tactile Interaction with Two-Person Dual-Coil fMRI," *Frontiers in Psychiatry* 11 (2020): article 279.

172 Cortisol levels, which remarkably tend to align: T. Field, "Romantic Breakups, Heartbreak and Bereavement," *Psychology* 2, no. 4 (2011): 382–87.

173 Rebounders study: Claudia C. Brumbaugh and R. Chris Fraley, "Too Fast, Too Soon? An Empirical Investigation into Rebound Relationships," *Journal of Social and Personal Relationships* 32 (2015): 99–118.

CHAPTER 15: EXCUSE MY PILOERECTION: THE SCIENCE OF AWE

177 For more on Simone Weil's wisdom, see Eric Weiner, *The Socrates Express* (New York: Avid Reader Press, 2020).

178 People behave in more generous ways after viewing epic nature: J. W. Zhang et al., "An Occasion for Unselfing: Beautiful Nature Leads to Prosociality," *Journal of Environmental Psychology* 37 (2014): 61–72. And Melanie Rudd, Kathleen D. Vohs, and Jennifer Aaker, "Awe Expands People's Perception of Time, Alters Decision Making, and Enhances Well-Being," *Psychological Science* 23 (2012): 1130–36.

179 The T-rex study and self-appraisal studies are both described in Michelle N. Shiota, Dacher Keltner, and Amanda Mossman, "The Nature of Awe: Elicitors, Appraisals, and Effects on Self-Concept," *Cognition and Emotion* 21, no. 5 (2007): 944–63.

179 The studies suggesting that people may feel literally smaller: M. van Elk et al., " 'Standing in Awe': The Effects of Awe on Body Perception and the Relation with Absorption," *Collabra* 2, no. 1 (2016): 4; and M. van Elk et al., "The Neural Correlates of the Awe Experience: Reduced Default Mode Network Activity during Feelings of Awe," *Human Brain Mapping* 40 (2019): 3561–74.

179 Yosemite study: Y. Bai et al., "Awe, the Diminished Self, and Collective Engagement: Universals and Cultural Variations in the Small Self," *Journal of Personality and Social Psychology* 113, no. 2 (2017): 185–209.

179 Calming down in the face of awe: A. M. Gordon et al., "The Dark Side of the Sublime: Distinguishing a Threat-Based Variant of Awe," *Journal of Personality and Social Psychology* 113, no. 2 (2016): 310–28.

180 The IL-6 study: J. E. Stellar et al., "Positive Affect and Markers of Inflammation: Discrete Positive Emotions Predict Lower Levels of Inflammatory Cytokines," *Emotion*, published online January 19, 2015.

CHAPTER 16: SPLIT MOUNTAIN

187   He and colleagues tested backpackers: Ruth Ann Atchley et al., "Creativity in the Wild: Improving Creative Reasoning through Immersion in Natural Settings," *PloS One* 7, no. 12 (2012): e51474.

187   canoeists in the Boundary Waters: Frank M. Ferraro III, "Enhancement of Convergent Creativity Following a Multiday Wilderness Experience," *Ecopsychology* 7 (2015): 7–11.

187   The backpacking study finding no mere vacation effect on cognition: Terry Hartig, Marlis Mang, and Gary W. Evans, "Restorative Effects of Natural Environment Experiences," *Environment and Behavior* 23 (1991): 3–26.

187   The river-running study looking at awe: Craig L. Anderson, Maria Monroy, and Dacher Keltner, "Awe in Nature Heals: Evidence from Military Veterans, At-Risk Youth, and College Students," *Emotion* 18, no. 8 (2018): 1195.

187   Circadian study: Ellen R. Stothard et al., "Circadian Entrainment to the Natural Light-Dark Cycle across Seasons and the Weekend," *Current Biology* 27, no. 4 (2017): 508–13.

188   In a study of 700 wilderness-therapy participants: Robert Greenway, "The Wilderness Effect and Ecopsychology," in *Ecopsychology*, ed. Theodore Roszak et al. (San Francisco: Sierra Club Books, 1995), 122–35.

189   Powell kept a short, scrappy journal on the river, but embellished it mightily half a dozen years later for the published version. The canyon quote is from Edward Dolnick, *Down the Great Unknown: John Wesley Powell's 1869 Journey of Discovery and Tragedy through the Grand Canyon* (New York: HarperCollins, 2001), Kindle.

191   Buzz saw quote: Wallace Stegner, *Beyond the Hundreth Meridian* (Lincoln: University of Nebraska Press, 1982), 20.

191   The loss of the No Name: Stegner, *Beyond the Hundreth Meridian*, 63–64.

191   Bradley quote: Stegner, 66.

192   Edward Abbey on vicious loveliness: Edward Abbey, "Floating," in *The Best of Edward Abbey* (San Francisco: Sierra Club Books, 2005), 349.

193   Survey of teens on bravery: "Dare to Dream, Dare to Act: What Girls Say about Bravery," sponsored by the Keds Brave Life Project, and conducted by Fluent, Girls Leadership Institute, 2014, https://girlsleadership.org/app/uploads/2014/11/BraveryBriefReport.pdf.

193   The REI survey was conducted by Edelman Intelligence in 2017: "REI's 2017 National Study on Women and the Outdoors," REI, March 30, 2017, https://www.slideshare.net/REI_/2017-national-study-on-women-and-the-outdoors.

196   The lottery system for the Green River: The federal permitting agency, the Bureau of Land Management, now lets you transfer a permit you win to another person if you can't make it.

197   and a $50 tribal fee: Six months after our trip, the Ute Indian Tribe of the Uintah and Ouray Reservation closed their land on the left bank for camping and hiking by non-members of the tribe, citing some disrespectful river runners who stole antlers from the reservation and didn't pay permit fees.

200   Mary Oliver quote from her essay "Of Power and Time," in *Upstream: Selected Essays*

(New York: Penguin, 2016); thanks to Brainpickings.org for highlighting the quote in the October 12, 2016, edition.

201 For a great overview of Desolation's natural and cultural resources, see James Aton, *The River Knows Everything* (Logan: Utah State University Press, 2009).

201 The basin has lost 19 percent of its water: Bradley Udall and Jonathan Overpeck, "The Twenty-First Century Colorado River Hot Drought and Implications for the Future," *Water Resources Research* 53, no. 3 (2017): 2404–18.

202 Meloy quote: Ellen Meloy, *Raven's Exile: A Season on the Green River* (New York: Henry Holt, 1994), 9.

CHAPTER 17: CONFLUENCE

203 Buzz Holmstrom: Quoted in Vince Welch et al., *The Doing of the Thing* (Flagstaff, AZ: Fretwater Press, 2004), 99.

204 Edward Abbey on the loneliest of primitive regions: From "Down the River with Henry Thoreau," *The Best of Edward Abbey* (San Francisco: Sierra Club Books, 2005), 272.

208 On goosenecks, Powell is quoted in Edward Dolnick, *Down the Great Unknown: John Wesley Powell's 1869 Journey of Discovery and Tragedy through the Grand Canyon* (New York: HarperCollins, 2001), Kindle.

209 For more on Victor Turner, see his *The Ritual Process: Structure and Anti-Structure* (Oxfordshire, UK: Routledge Press, 1995).

210 Wilderness as a "retreat from cultural dominance": Thomas Doherty, "Robert Greenway: The Ecopsychology Interview," *Ecopsychology* (March 2009): 47–52.

210 "No one ever told me that grief felt so like fear": C. S. Lewis, *A Grief Observed* (New York: HarperCollins, 1996), 15.

212 one of their current studies suggests that we may be processing external rewards differently: This paper hadn't been accepted at time of publication, but it's worth checking out Emily Scott and Sara LoTemplio's other studies looking at brain patterns in nature: R. J. Hopman et al., "Resting-State Posterior Alpha Power Changes with Prolonged Exposure in a Natural Environment," *Cognitive Research: Principles and Implications* 5, no.1 (2020): 1–24; S. B. LoTemplio et al., "Nature as a Potential Modulator of the Error-Related Negativity: A Registered Report," *International Journal of Psychophysiology* (June 14, 2020).

CHAPTER 18: THE HAPPINESS THAT MATTERS: SOCIAL WELL-BEING

221 For more on hedonic versus eudaimonic happiness, see C. D. Rhyff, "Self-Realisation and Meaning Making in the Face of Adversity: A Eudaimonic Approach to Human Resilience," *Journal of Psychology in Africa* 24, no. 1 (2014); and R. M. Ryan and E. L. Deci, "On Happiness and Human Potentials: A Review of Research on Hedonic and Eudaimonic Well-Being," *Annual Review of Psychology* 52 (2001): 141–66.

221 Fredrickson and Cole's 2013 study: Barbara L. Fredrickson et al., "A Functional Genomic Perspective on Human Well-Being," *Proceedings of the National Academy of Sciences* 110, no. 33 (August 13, 2013): 13684–89.

221   Japanese workers study: S. Kitayama et al., "Work, Meaning, and Gene Regulation: Findings from a Japanese Information Technology Firm," *Psychoneuroendocrinology* 72 (2016): 175–81.

222   Cole's retirement study: S. W. Cole et al., "Loneliness, Eudaimonia, and the Human Conserved Transcriptional Response to Adversity," *Psychoneuroendocrinology* 62 (2015): 11–17.

222   For more on the impacts of prosociality on gene expression, see: S. K. Nelson-Coffey et al., "Kindness in the Blood: A Randomized Controlled Trial of the Gene Regulatory Impact of Prosocial Behavior," *Psychoneuroendocrinology* 81 (2017).

223   The Generation Xchange study: T. Seeman et al., "Intergenerational Mentoring, Eudaimonic Well-Being and Gene Regulation in Older Adults: A Pilot Study," *Psychoneuroendocrinology* 111 (2020).

CHAPTER 19: TRUTH SERUM, PART TWO

227   For more on cognitive closure, see Michelle N. Shiota, Dacher Keltner, and Amanda Mossman, "The Nature of Awe: Elicitors, Appraisals, and Effects on Self-Concept," *Cognition and Emotion* 21, no. 5 (2007): 944–63; and D. M. Webster and A. W. Kruglanski, "Individual Differences in Need for Cognitive Closure," *Journal of Personality and Social Psychology* 67, no. 6 (1994): 1049–62.

228   The Keltner-Haidt paper: D. Keltner and J. Haidt, "Approaching Awe, a Moral, Spiritual, and Aesthetic Emotion," *Cognition and Emotion* 17, no. 2 (2003): 297–314.

228   Keltner's "bursts of passion" quote is from his book *Born to Be Good: The Science of a Meaningful Life* (New York: W. W. Norton, 2009), ix. All other quotes are from personal communication.

CHAPTER 20: THE DIVORCE DRUG

232   MDMA's effects on kidneys: Feyza Bora, Fatih Yilmaz, and Taner Bora, "Ecstasy (MDMA) and Its Effects on Kidneys and Their Treatment: A Review," *Iranian Journal of Basic Medical Sciences* 19 (2016): 1151–58.

232   MDMA is known to cause a cranial release of neurohormones: See G. J. Dumont et al., "Increased Oxytocin Concentrations and Prosocial Feelings in Humans after Ecstasy (3,4-Methylenedioxymethamphetamine) Administration," *Social Neuroscience* 4 (2009): 359–66; and C. M. Hysek, G. Domes, and M. E. Liechti, "MDMA Enhances 'Mind Reading' of Positive Emotions and Impairs 'Mind Reading' of Negative Emotions," *Psychopharmacology* 222 (2012): 293–302.

233   Pollan's book: Michael Pollan, *How to Change Your Mind: What the New Science of Psychedelics Teaches Us about Consciousness, Dying, Addiction, Depression, and Transcendence* (New York: Penguin, 2019).

CHAPTER 21: OPEN SESAME

241   at least one scientist believes that even one hit of MDMA can permanently damage serotonin pathways: This was reported in Matt Klam's *New York Times Magazine* story "Experiencing Ecstasy," January 21, 2001.

242   I was reminded of a line by British journalist and naturalist Michael McCarthy: From Michael McCarthy, *The Moth Snowstorm* (New York: New York Review of Books, 2016), chap. 2.

242   one Australian study showed that a one-deviation jump in openness: C. J. Boyce, A. M. Wood, and N. Powdthavee, "Is Personality Fixed? Personality Changes as Much as 'Variable' Economic Factors and More Strongly Predicts Changes to Life Satisfaction," *Social Indicators Research* 111 (2013): 287–305.

243   one of the first studies demonstrating changes in personality: Katherine A. MacLean, Matthew W. Johnson, and Roland R. Griffiths, "Mystical Experiences Occasioned by the Hallucinogen Psilocybin Lead to Increases in the Personality Domain of Openness," *Journal of Psychopharmacology* 25, no. 11 (2011): 1453–61.

243   In one study from 2017: Matthew M. Nour, Lisa Evans, and Robin L. Carhart-Harris, "Psychedelics, Personality and Political Perspectives," *Journal of Psychoactive Drugs* 49, no. 3 (2017): 182–91.

243   University of South Carolina study on openness: M. T. Wagner et al., "Therapeutic Effect of Increased Openness: Investigating Mechanism of Action in MDMA-Assisted Psychotherapy," *Journal of Psychopharmacology* 31, no. 8 (2017): 967–74.

244   The MEQ revises a questionnaire from the 1960s: K. A. Maclean et al., "Factor Analysis of the Mystical Experience Questionnaire: A Study of Experiences Occasioned by the Hallucinogen Psilocybin," *Journal for the Scientific Study of Religion* 51, no. 4 (2012): 721–37.

245   Hildegard's vision: Barbara Newman, "Hildegard of Bingen: Visions and Validation," *Church History* 54, no. 2 (1985): 163–75.

245   Hendricks paper: Peter Hendricks, "Awe: A Putative Mechanism Underlying the Effects of Classic Psychedelic-Assisted Psychotherapy," *International Review of Psychiatry* 30, no. 4 (2018): 331–42.

246   Abraham Maslow, the twentieth-century psychologist who conceptualized: A good overview of the past and updated versions of the hierarchy of needs can be found in M. Koltko-Rivera, "Rediscovering the Later Version of Maslow's Hierarchy of Needs: Self-Transcendence and Opportunities for Theory, Research, and Unification," *Review of General Psychology* 10, no. 4 (2006): 302–17.

247   The Marsh Chapel Experiment researchers were taking psychedelics: Griffiths describes this at amusing length in Sam Harris's *Making Sense* podcast, no. 177, December 2, 2019.

247   Griffiths's 2002 experiment was published four years later: R. R. Griffiths et al., "Psilocybin Can Occasion Mystical-Type Experiences Having Substantial and Sustained Personal Meaning and Spiritual Significance," *Psychopharmacology* (2006): 187–268.

247   Maslow as a natural mystic: Bill Richards discusses Maslow's mysticism in William A. Richards, "Abraham Maslow's Interest in Psychedelic Research: A Tribute," *Journal of Humanistic Psychology* 57, no. 4 (2017): 319–22.

247   Other studies have found shroom-induced deactivation in the posterior cingulate cortex: Robin L. Carhart-Harris et al., "Neural Correlates of the Psychedelic State as

Determined by fMRI Studies with Psilocybin," *Proceedings of the National Academy of Sciences* 100, no. 15 (2012): 8788–92.

248   The drugs disrupt the normal pathways: see A. V. Lebedev et al., "Finding the Self by Losing the Self: Neural Correlates of Ego-Dissolution under Psilocybin," *Human Brain Mapping* 36, no. 8 (2015): 3137–53.

248   For more info about the pathways of the salience network, see V. Menon, "Salience Network," *Brain Mapping: An Encyclopedic Reference*, ed. Arthur W. Toga (Cambridge, MA: Academic Press, 2015), 2:597–611, https://med.stanford.edu/content/dam/sm/scsnl/documents/Menon_Salience_Network_15.pdf.

248   For brain structures making up the DMN, see Marcus Raichle, "The Brain's Default Mode Network," *Annual Review of Neuroscience* 38 (2015): 433–47.

248   For Iris Murdoch's description of seeing the kestrel, I am indebted to Maria Popova's Brain Pickings, October 21, 2019, edition, https://www.brainpickings.org/2019/10/21/iris-murdoch-unselfing/.

249   For a good discussion of Freud's use of the term *decathexis*, see Tammy Clewell, "Mourning beyond Melancholia: Freud's Psychoanalysis of Loss," *Journal of the American Psychoanalytic Association* 52, no. 1 (2004): 43–67.

CHAPTER 22: MAN IN THE KASTLE: OPIOIDS, LOVE, AND THE SCIENCE OF RECOVERY

252   For DeWall's Tylenol study, see C. N. DeWall et al., "Acetaminophen Reduces Social Pain," *Psychological Science* 21, no. 7 (2010): 931–37.

252   DeWall's quoted marijuana study: T. Deckman et al., "Can Marijuana Reduce Social Pain?," *Social Psychological and Personality Science* 5, no. 2 (2014): 131–39.

255   Panksepp quotes from Jaak Panksepp, *Affective Neuroscience: The Foundations of Human and Animal Emotions* (Oxford: Oxford University Press, 1998), 248. Homer passage cited on page 264.

258   According to the National Institutes of Health, at least two years of treatment is necessary to maintain long-term recovery: David Uhl and David Coffey, *Alternative Therapeutic Support Community Model in Seattle and Expanding Alternative Therapeutic Community Model to New Locations in King County*, MIDD Briefing Paper (2015), 3.

CHAPTER 24: THE PERSONALITY OF THE BODY

268   we consulted the wiki of genes: We used Genecards.org, an open-source compendium that geneticists contribute to as they learn more about the human genome and the functions of our 20,000 genes.